FOUNDATIONS AND RETAINING STRUCTURES

Foundations and Retaining Structures

Editor:

Dr. Tanjina Nur, Ph.D.

ArclerPress

www.arclerpress.com

Foundations and Retaining Structures
Editor: Dr. Tanjina Nur, Ph.D.

708 3rd Avenue, 6th Floor
New York
NY 10017
United States of America
www.arclerpress.com

ISBN: 978-1-68094-375-7

Library of Congress Control Number: 2016950953

About the Editor

Dr. Tanjina Nur, Ph.D.

Tanjina finished her PhD in Civil and Environmental Engineering in 2014 from University of Technology Sydney (UTS). Now she is working as Post-Doctoral Researcher in the Centre for Technology in Water and Wastewater (CTWW) and published about eight International journal papers with 80 citations. Her research interest is wastewater treatment technology using adsorption process.

Contents

List of Contributors

Nabeel A. Jasim
Department of Civil Engineering, College of Engineering, University of Basra, Basra, Iraq

Ahmed M. Al-Yaqoobi
Department of Civil Engineering, College of Engineering, University of Basra, Basra, Iraq

Kwangkuk Ahn
Department of Civil Engineering, Chungbuk National University, Cheongju, Chungbuk 361-763, Republic of Korea

Hongsig Kang
Department of Civil Engineering, Chungbuk National University, Cheongju, Chungbuk 361-763, Republic of Korea

Hong-zhi Qiu
Key Laboratory of Mountain Hazards and Earth Surface Processes, Institute of Mountain Hazards and Environment, Chinese Academy of Sciences, Chengdu, Sichuan 610041, China
University of Chinese Academy of Science, Beijing 100049, China

Ji-ming Kong
Key Laboratory of Mountain Hazards and Earth Surface Processes, Institute of Mountain Hazards and Environment, Chinese Academy of Sciences, Chengdu, Sichuan 610041, China

Ren-chao Wang
Key Laboratory of Mountain Hazards and Earth Surface Processes, Institute of Mountain Hazards and Environment, Chinese Academy of Sciences, Chengdu, Sichuan 610041, China
University of Chinese Academy of Science, Beijing 100049, China

Dong Li, Wei Wang
Tianjin Key Laboratory of Nonlinear Dynamics and Chaos Control, School of Mechanical, Tianjin University, Tianjin 300072, China

Qichang Zhang
Tianjin Key Laboratory of Nonlinear Dynamics and Chaos Control, School of Mechanical, Tianjin University, Tianjin 300072, China

Yang Changwei
School of Civil Engineering, Key Laboratory of Transportation Tunnel Engineering, Ministry of Education, Southwest Jiaotong University, Chengdu 610031, China

Zhang Shixian
School of Civil Engineering, Key Laboratory of Transportation Tunnel Engineering, Ministry of Education, Southwest Jiaotong University, Chengdu 610031, China
School of Civil Engineering, Panzhihua University, Panzhihua 617000, China

Zhang Jianjing
School of Civil Engineering, Key Laboratory of Transportation Tunnel Engineering, Ministry of Education, Southwest Jiaotong University, Chengdu 610031, China

Bi Junwei
School of Civil Engineering, Key Laboratory of Transportation Tunnel Engineering, Ministry of Education, Southwest Jiaotong University, Chengdu 610031, China

Changwei Yang
School of Civil Engineering, Key Laboratory of Transportation Tunnel Engineering, Ministry of Education, Southwest Jiaotong University, Chengdu 610031, China

Jian Jing Zhang
School of Civil Engineering, Key Laboratory of Transportation Tunnel Engineering, Ministry of Education, Southwest Jiaotong University, Chengdu 610031, China

Qu Honglue
School of Civil Engineering, Key Laboratory of Transportation Tunnel Engineering, Ministry of Education, Southwest Jiaotong University, Chengdu 610031, China

Bi Junwei
School of Civil Engineering, Key Laboratory of Transportation Tunnel Engineering, Ministry of Education, Southwest Jiaotong University, Chengdu 610031, China

Liu Feicheng
School of Civil Engineering, Key Laboratory of Transportation Tunnel Engineering, Ministry of Education, Southwest Jiaotong University, Chengdu 610031, China

Seyed Amin Mousavi
School of Civil Engineering, University of Tehran, Tehran, Iran

Khosrow Bargi
School of Civil Engineering, University of Tehran, Tehran, Iran

S. Talatahari
Marand Faculty of Engineering, University of Tabriz, Tabriz, Iran

R. Sheikholeslami
Department of Civil and Environmental Engineering, Amirkabir University of Technology, Tehran, Iran

M. Shadfaran
Department of Civil Engineering, University of Tabriz, Tabriz, Iran

M. Pourbaba
Department of Civil Engineering, Islamic Azad University, Maragheh Branch, Maragheh, Iran

D. M. Dewaikar
Department of Civil Engineering, Indian Institute of Technology Bombay, Mumbai 400076, India

S. R. Pandey
Department of Civil Engineering, Indian Institute of Technology Bombay, Mumbai 400076, India

Jagabandhu Dixit
Department of Civil Engineering, Indian Institute of Technology Bombay, Mumbai 400076, India

Sreenivas Alampalli
Transportation Research & Development Bureau, New York State Department of Transportation, 1220, Washington Avenue, Albany, NY 12232-0869, USA

Ahmed-W. Elgamal
University of California at San Diego, San Diego, CA 92093-0085, USA

A. Sekkel
Moulay Tahar University Dept. Civil Engineering & Hydraulic Fac. Technology Labo. (L.G.C.E) Saida, Algeria

M. Meghachou
Djillali Liabes University Dept. Civil Engineering Fac. Technology Labo. (L.G.C.E) Sidi Bel-Abbes, Algeria

Amina Lazizi
Civil Engineering Dpt, Djillali Liabes University of Sidi Bel Abbes, Algeria

Habib Trouzine
LGCE Laboratory, Djillali Liabes University of Sidi Bel Abbes, Algeria

Aissa Asroun
LGCE Laboratory, Djillali Liabes University of Sidi Bel Abbes, Algeria

Farid Belabdelouahab
ENSTP National High School of Publics Work, Kouba, Algiers, Algeria

Behzad Nematollahi
Department of Civil Engineering, Faculty of Engineering, Universiti Putra Malaysia (UPM), 43400 UPM, Serdang, Malaysia

M.R. Raizal Saifulnaz
Department of Civil Engineering, Faculty of Engineering, Universiti Putra Malaysia (UPM), 43400 UPM, Serdang, Malaysia

Yen Lei Voo
Dura Technology Sdn. Bhd., Perak, Ipoh, Malaysia

Shodolapo Oluyemi Franklin
Department of Civil Engineering, University of Botswana, P. Bag 0061, Gaborone, Botswana

Abiodun Luqman Olopade
Formerly, Department of Civil Engineering, University of Ibadan, Ibadan, Nigeria

Shodolapo Oluyemi Franklin
Department of Civil Engineering, University of Botswana, P. Bag 0061, Gaborone, Botswana

Abiodun Luqman Olopade
Formerly, Department of Civil Engineering, University of Ibadan, Ibadan, Nigeria

Venkata Ramasubbarao Godavarthi
Department of Civil Engineering, SRK Institute of Technology, Enikepadu-521 108, Vijayawada, Andhra Pradesh, India

Dineshbabu Mallavalli
Department of Civil Engineering, SRK Institute of Technology, Enikepadu-521 108, Vijayawada, Andhra Pradesh, India

Ramya Peddi
Department of Civil Engineering, SRK Institute of Technology, Enikepadu-521 108, Vijayawada, Andhra Pradesh, India

Neelesh Katragadda
Department of Civil Engineering, SRK Institute of Technology, Enikepadu-521 108, Vijayawada, Andhra Pradesh, India

Prudhvikrishna Mulpuru
Department of Civil Engineering, SRK Institute of Technology, Enikepadu-521 108, Vijayawada, Andhra Pradesh, India

Reza Dezvareh
School of Civil Engineering, College of Engineering, University of Tehran, Iran

Prof. Khosrow Bargi
School of Civil Engineering, College of Engineering, University of Tehran, Iran

Yasin Moradi
School of Civil Engineering, College of Engineering, University of Tehran, Iran

Preface

This book covers a broad range of review of retaining structures of foundations. Retaining structures are engineered to retain soil and/or rock. There are several types of retaining structures, including retaining walls, gravity, cantilever, sheet pile, and anchored earth and mechanically stabilized earth (reinforced earth) walls. Firstly, we discuss optimum deign of tied back retaining walls and behavior of Reinforced Retaining Walls with Different Reinforcement Spacing during Vehicle Collisions. Then Dynamic Active Earth Pressures of the Retaining Piles with Anchors and Lateral Earth Pressure behind Walls Rotating about Base considering Arching Effects are presented in details. We also try to find out seismic stability time-frequency analysis method of reinforced retaining wall and seismic Earth pressures of retaining wall from large shaking table tests.

In the middle part of the book, we focus on finding enhancing seismic capacity of pile-supported wharves using yielding dampers and optimum design of gravity retaining walls using charged system search algorithm. Active thrust on an inclined retaining wall with inclined cohesion less backfill due to surcharge effect and dynamic response of wall backfill retaining system are also discussed. Then we present the effect of displacement mode of rigid retaining walls on shearing bands by active earth pressure and numerical simulation of tire reinforced sand behind retaining wall under earthquake excitation. At the end, we try to describe sustainability assessment of precast ultra-high Performance fiber reinforced concrete (UHPFRC) cantilever retaining walls. Also A note and some remarks on the analysis and design of cantilever retaining walls and anchored sheet pile walls using the CECP2 and BS8002 methods are also discussed. Then contiguous pile wall as a deep excavation supporting system and incremental dynamic analysis of sheet pile quay-walls in order to reach performance-based design criteria for this type of wharf are presented briefly

Editor

Dr. Tanjina Nur, Ph.D.

INTRODUCTION

A foundation (or, more commonly, foundations) is the element of an architectural structure which connects it to the ground, and transfers loads from the structure to the ground. Foundations are generally considered either shallow or deep. Foundation engineering is the application of soil mechanics and rock mechanics (Geotechnical engineering) in the design of foundation elements of structures.

Whenever construction workers begin work on a new building, they must first assess where and how they will build the foundation. This is a structure, commonly made of concrete for homes that transfers the weight of the building onto the earth below. There are different types of foundation designs and each serves a different specific purpose, but generally, every one works to transfer the weight load of a structure to the soil beneath.

Most small and medium homes are built upon a shallow foundation. These are usually comprised of concrete strips that are laid about 3.3 feet (1 meter) beneath the soil, or of a single large concrete slab that is also set about 3.3 feet (1 meter) beneath the soil. When applicable, it will extend beneath the frost line. When engineers design the foundation of a building, they must keep in mind how much the soil will settle beneath it, as well as how much weight will go on top. If calculated incorrectly, the foundation may fail and place the entire structure in peril.

Larger buildings use a deep rather than a shallow foundation. A deep foundation uses long pylons of steel or concrete to penetrate beyond the weaker surface soils into the deeper and more stable soils or bedrock beneath. The load from the walls above is transferred deep into the earth, thereby providing support for the intense weight above. Much like the considerations for a shallow foundation, engineers must consider weight and settlement, as well as scour — water eroding soil beneath the structure.

For structures being built in colder climates, engineers must consider frost heaves as well. Frost heaves occur when moisture in the soil freezes, thereby changing the density of the building's support. A frost heave can cause damage to the foundation, thereby compromising the structural integrity of the entire building. Drier, warmer climates are not entirely exempt from such worries, however; certain soils will expand and contract when moisture is added or taken away, and engineers must factor in such movement when considering where and how to lay a foundation.

Raised Foundation

A raised foundation is a foundation that, as its name would suggest, is the main floor of a home or business that is raised above the plane of the surrounding earth. Along with the concrete slab foundation, the raised version accounts for most of the foundation options used for homes in most of the world at least the industrialized world. These foundations offer some benefits, but also has some disadvantages.

Many things can be used to raise the foundation above the ground. Concrete blocks or poured concrete forms are often used to raise the foundation. Often around the outside of the home, the outer edges of the support are concrete, brick, or some other solid material. After the pylons are put into place, the next step is to bridge the pylons with another material, usually wood.

Many say that the aesthetics of a raised home are far superior to that of slab homes. In a neighborhood of slab homes, a home with a raised foundation usually stands out and is often a little taller that most of the homes around it. Still, when it comes down to aesthetics, this is more of a personal preference than a true fact.

Despite that, there are some true advantages to homes with a raised foundation. For example, there is less of a need for grading and leveling if a home has such a foundation. Also, in a flood-prone area, there is some additional protection against flooding.

Another benefit of a raised foundation comes when working under the home. Getting to plumbing or other areas under the home is easier with a raised foundation. Also, this type of foundation allows for the opportunity to put in a basement.

However, a number of disadvantages are also associated with a raised foundation. The maintenance required on them is usually more than is required for concrete slab homes. Also, the crawl space underneath the home provides an ideal living space for a number of undesirable animals. While there may be ways to keep such animals out, it may be a constant battle.

It is difficult to say whether a home with a raised foundation is a more difficult task to handle during home construction than building one with a slab foundation. In some ways, this may depend on the location. At a spot with level ground and compacted earth, slab homes may not be very much of a problem at all. Each type of construction presents its own unique opportunities and challenges.

Floating Foundation

A floating foundation is the foundation of a building that does not use footings. It is a poured cement slab that has two deep edges that go just below the frost line in northern climates. The foundation actually does float on the earth and moves as temperatures compact and expand the soil. It is very common in garage floors and mobile home slabs, and in many parts of the world, it's simply referred to as a slab foundation.

In a typical floating foundation, the plumbing and electrical lines are fastened to the slab by simply running them through the floor as it is poured. This means that the plumbing, drainage, and electrical lines must all be completed prior to pouring the foundation. In many cases, it is a much more affordable method of building than using a footing equipped foundation, but it can lead to very expensive repair bills if the plumbing requires work in the future. In warmer climates, the floating slab is much friendlier to homeowners as it does not flex too much because there are fewer extreme changes in temperature. It also helps to cool the structure, as the foundation's contact with the ground draws cool temperatures through the concrete and disperses them throughout the building.

When constructing out-buildings such as sheds, the term can take on another meaning. Some flooring systems for outside sheds are known as floating foundation floors and do not require any concrete to be poured. In these applications, the foundation is merely lumber framing placed upon blocks. This allows the floor of the shed to sit elevated off of the ground, which prevents water from seeping into the shed and protects the shed's contents.

When pouring a floating slab foundation, it is imperative to install reinforced steel rods or heavy wire mesh in the floor before pouring the concrete. The wire and steel prevent the floor from cracking and breaking as the slab flexes with the earth. Without this reinforcement, the slab will likely crack and possibly cause damage to the walls and ceiling, as the floor will be allowed to flex unevenly. Often, the flexing of an unreinforced slab will cause doors to not close or open correctly and windows to stick or even crack. Walls are also subject to cracking, and drywall is prone to break at the seams.

Slab Foundation

A slab foundation is a layer of concrete, generally poured out over a prepared surface of soil or gravel, upon which a house or other structure is built. More appropriate to climates where ground freezing and thawing, and the associated soil movement, are not of particular concern, slab foundations are nevertheless anchored into the ground securely to prevent structural damage from environmental causes such as severe weather or shifting of the soil. While a structure built on a slab foundation has no "below grade" space that can be used for residential or storage purposes, it's generally much less costly to build than a comparable structure with a basement.

Besides providing a stable, flat, and level surface for occupants, a slab's purpose is to distribute much of the weight that's placed upon it. For example, many walls are "load bearing," and without a foundation of some sort to rest upon, would not only sink into the soil to some extent, but would also respond to the soil's every movement. This is one reason that the soil underneath slab-type foundations must be prepared before the concrete is poured. For example, the soil will usually be leveled and overlaid with gravel and moisture barriers. The thickness of the foundation is calculated based upon a number of factors, including the anticipated load the slab will bear.

Another critical element of slab construction is footings — "anchors" of concrete poured into holes dug in the ground for that purpose. When the foundation is poured, it either rests on these anchors or is attached to them by means of rebar, reinforcing bars made of steel that are sunk into the footings while they're still wet, providing a high degree of stability for the slab foundation.

The main benefit of slab-type foundations is cost; a slab is the least costly foundation option available. There are many drawbacks to a slab foundation, though, especially in residential construction. For instance, a slab puts a great deal of stress on the legs, feet, and backs of the inhabitants when they walk on the ground floor. Even with carpeting, much of the shock of every step taken is absorbed by the skeleton and not the floor. Heating costs in slab-foundation houses are generally higher also, because so much cold comes into the house through the floor. Plumbing and electrical lines are either run down through the slab, making access difficult, or must be run out a wall, incurring added costs of securing and weatherizing. Another significant drawback of a slab foundation is that space must be given over to appliances that would normally be maintained in a basement, such as hot water heaters, furnaces, and HVAC equipment.

Mat Foundation

A mat foundation is a type of shallow foundation made by pouring concrete over a reinforcing mat of material, usually rebar. This design is also known as a raft foundation in some regions. A mat foundation distributes the weight of the structure very close to the surface, rather than deep into the ground, and it is suitable for a range of environments. If an engineer thinks that a mat foundation is appropriate or necessary for a structure, he or she can design the specifications to meet the need and can supervise the creation of the mat and the pouring of the foundation.

The depth of this style of foundation can vary and will depend on the specifications of the project. Engineers can find mat foundations helpful when they need space under a structure for underground facilities or when they build on potentially unstable soils. The foundation starts with a tight mat made from rebar to create a lattice of support. Some designs include plastic blocks for weight distribution that will also cut down on the amount of concrete needed.

Construction workers might require several days to set up and secure all of the rebar. Tensioner bars can be placed inside the mat to keep it stable. The workers pour concrete continuously over the foundation until the mat is entirely covered, and they use tools to vibrate it into place to eliminate air bubbles and any other issues that arise, before finishing off the top and allowing the foundation to set. After the foundation is finished curing, building can progress.

The finished mat foundation should distribute the weight evenly, without specific load-bearing pillars that send the weight of the structure deep underground. It can cover a large area and will remain stable and solid in a variety of conditions. Mat foundations can be used for residential and commercial construction. Foundation work such as repairs can be challenging with this design because it is a flat, closed area of concrete without easy access points.

Engineers and foundation specialists typically are needed to design a mat foundation unless the area is very small, such as a foundation under a shed. They review the work site and the building specifications to determine what kinds of loads will be present and how the soil under the building might behave. This information is valuable for the stage of the process when they design the foundation, including the configuration of supports in the mat to distribute weight and maintain tension across the finished foundation.

Foundation Wall

A foundation wall is a wall which rests on the footers of the foundation, and provides support for the weight of a house. Foundation walls are a critical part of the construction of a home with a perimeter foundation design, and they are one of the areas of construction in which it most definitely pays to take the time to do the job right. Problems with the foundation wall will ultimately lead to problems with the house, up to and including complete structural failure.

A number of materials can be used to make a foundation wall. Poured concrete is a common and popular choice, with some contractors using modular walls which are actually poured off site and transported to a building site. Blocks, bricks, and stone can also be used. In all cases, the materials must be load bearing, and the wall needs to be designed and built so that the weight will be distributed evenly over the footers of the house.

One of the most common problems with foundation walls is cracking. Foundation cracks can be benign or serious, depending on the location of the crack, the size, and the direction of the crack. Cracks can appear because the wall was built in a hurry, backfilled before the concrete had a chance to cure, or not properly reinforced. Earthquakes and settling of the underlying ground can also result in cracking of the foundation, as can damage caused by impact to the foundation.

Foundation walls can be installed after a house is built, as seen when a house with a post and pier foundation is retrofitted to install a perimeter foundation. Likewise, damaged foundations can also be replaced. In both cases, the house is usually raised on jacks which support the weight of the house while the foundation work is done. Foundation stabilization can be tricky work, and in urban areas, there are usually entire contracting firms dedicated specifically to repairing, stabilizing, and reinforcing foundations.

For commercial structures, building foundation walls can become much more complicated. The walls must be much larger, much more heavily reinforced, and capable of bearing huge amounts of weight once the building is finished. Engineering firms are usually consulted to design the foundation wall and determine the load requirements, ensuring that the wall will ideally exceed the need, as replacing foundations as a result of a foundation wall failure on large structures is not an enjoyable or cheap process.

Retaining Wall

A retaining wall is a stabilizing structure used to hold sloping ground in place and to prevent the erosion and the movement of soil. It must be substantial and sturdy enough to accommodate and redistribute the lateral pressure caused by sloping ground. It is normally designed with seepage holes, which allow collected ground water to escape. This releases the additional pressure created by accumulated water and helps keep the wall stable.

Several different materials can be used to construct a retaining wall. Stone and concrete are often used, and there are special blocks crafted from aggregate materials and light concrete that are designed specifically for this purpose. Some styles interlock, making assembly simpler, less costly and less time consuming, because each block fits securely with the next. Because the fit of these blocks is secure, they do not require the addition of mortar to hold them together.

A retaining wall can be a series of "steps" or tiers, which allows for a more attractive design as well as more efficient erosion control. The design can include different types of plantings, flowers or materials in each tier to bring more texture, color and interest to the area. Aside from the aesthetic value, a tiered design also provides better erosion control by breaking down the amount of soil, and therefore pressure, held by each division of the wall.

Large stones and railroad ties were often used to build a tiered retaining wall in the past. Today, there are many styles and types of blocks available for this purpose, and most people choose preformed blocks. They are affordable and easy to use, making other methods somewhat obsolete, unless the look of rough stone or wood is preferred. A homeowner can build the wall himself or have one built by a professional. The key to constructing a stable wall is making sure that it is completely level. Each layer must also be backfilled to give it substance and to help provide good drainage. Generally, sand, stone, pea gravel or leftover soil is used as backfill.

A retaining wall is a structure designed and constructed to resist the lateral pressure of soil when there is a desired change in ground elevation that exceeds the angle of repose of the soil.

A basement wall is thus one kind of retaining wall. But the term usually refers to a cantilever retaining wall, which is a freestanding structure without lateral support at its top. These are cantilevered from a footing and rise above the grade on one side to retain a higher level grade on the opposite side. The walls must resist the lateral pressures generated by loose soils or, in some cases, water pressures.

Every retaining wall supports a "wedge" of soil. The wedge is defined as the soil which extends beyond the failure plane of the soil type present at the wall site, and can be calculated once the soil friction angle is known. As the setback of the wall increases, the size of the sliding wedge is reduced. This reduction lowers the pressure on the retaining wall.

The most important consideration in proper design and installation of retaining walls is to recognize and counteract the tendency of the retained material to move downslope due to gravity. This creates lateral earth pressure behind the wall which depends on the angle of internal friction (phi) and the cohesive strength (c) of the retained material, as well as the direction and magnitude of movement the retaining structure undergoes.

Lateral earth pressures are zero at the top of the wall and – in homogenous ground – increase proportionally to a maximum value at the lowest depth. Earth pressures will push the wall forward or overturn it if not properly addressed. Also, any groundwater behind the wall that is not dissipated by a drainage system causes hydrostatic pressure on the wall. The total pressure or thrust may be assumed to act at one-third from the lowest depth for lengthwise stretches of uniform height.

Unless the wall is designed to retain water, It is important to have proper drainage behind the wall in order to limit the pressure to the wall's design value. Drainage materials will reduce or eliminate the hydrostatic pressure and improve the stability of the material behind the wall. Drystone retaining walls are normally self-draining

1

CHAPTER

OPTIMUM DESIGN OF TIED BACK RETAINING WALL

Nabeel A. Jasim, Ahmed M. Al-Yaqoobi

Keywords: Retaining Walls, Tied Back Wall, Optimization, Optimtool, Genetic Algorithm

INTRODUCTION

In design process, engineers have to take many technological and managerial decisions at several stages. The present design of economical concrete structures mainly follows rules based on the experience of structural engineers. Most procedures adopt the cross-section dimensions and material grades sanctioned by common practice. Structural optimization methods are clear alternatives to designs based on experience. Optimum design is a structural synthesis which collects all important engineering aspects to develop structural versions not only safe but also economic. The economy is achieved by minimizing a cost function and the safety is guaranteed by fulfilling the design constraints.

The constraints may be based on stability, bending moment and shear force capacities, and some of the other measures. Thus, the problem can be defined mathematically as a constrained function minimization task, which may be solved by a mathematical programming method [1] .

The improvements in numerical methods and computer technology have given impetus to this concept of optimization. In recent years, some optimization methods that are conceptually different from the traditional mathematical programming techniques have been developed. These methods are labeled as modern or nontraditional methods

Citation: Nabeel A. Jasim, Ahmed M. Al-Yaqoobi (2016) Optimum Design of Tied Back Retaining Wall. Open Journal of Civil Engineering,06,139-155. doi: 10.4236/ojce.2016.62013.

of optimization. Most of these methods are based on certain characteristics and behavior of biological, molecular, swarm of insects, and neurobiological systems.

Several authors have surveyed the utilization of optimization in structural design. Al-Janabi [2] (1983) studied the structural behavior of a proposed L-shape tied-back type of retaining walls. The finite element method was utilized to perform the calculations necessary to construct design curves for tie positions, which gives minimum values of bending moments. It should be pointed out that the study did not use any mathematical programming technique, but merely used the graphical representation of the results. Ceranic and Fryer [3] (1999) presented results of the application of a constrained simulated annealing algorithm to the minimum cost design of reinforced concrete cantilever retaining walls. A modified simulated annealing algorithm was proposed that avoids the simple rejection of infeasible solutions and improves convergence to a minimum cost. Results obtained so far have shown that simulated annealing can be successfully applied to the minimum cost design of reinforced concrete retaining walls. M. Ghazavi and A. Heidarpour [4] (2003) presented an optimization algorithm for the design of reinforced concrete counter fort retaining walls. A special efficient computer program was developed for this purpose. For the analysis of lateral earth pressures on the wall, the well known Coulomb, Rankine, and wedge methods were used. The backfill was treated as homogeneous or stratified also horizontal or inclined. The effects of surcharge, hydrostatic water pressure, seepage water pressure, and seismic loading on the lateral earth pressure were incorporated. Input parameters were generally the height of the wall, backfill slope, and backfill and base soil geotechnical parameters. Appropriate strengths for concrete and steel were introduced to the optimization scheme. Structural stability due to bending moment and shear force, geotechnical considerations such as sliding, overturning, settlement and bearing capacity were taken into account. The costs of construction material mainly concrete, reinforcing steel, and formwork had been considered for various acceptable materials. In the analysis, the geometry of the wall was optimized and compared with the recommended values. It was shown that the optimization of counter fort retaining walls can reduce the costs involved. Yepes et al. [5] (2007) examined the economic optimization of reinforced concrete earth-retaining walls used in road construction. The simulated annealing algorithm was the proposed method to optimize walls. The formulation of the problem included 20 design variables: four geometrical ones dealing with the thickness of the kerb and the footing, as well as the toe and the heel lengths; four material types; and 12 variables for the reinforcement set-up. The study estimated the relative importance of factors such as the base friction coefficient, the wall-fill friction angle and the limitation of kerb deflections. Finally, the paper presented a parametric study of commonly used walls from 4 to 10 m in height for different fills and bearing conditions. Average expressions were calculated for the total cost, the volume of concrete, the thickness of the kerb and the footing, the lengths of the footing and the heel, which may be useful for the practical design of walls. An upper bound of 50 kg/m^3 of reinforcement in the kerb and 60 kg/m^3 for the overall wall was reported. Magbo et al. [6] (2012) generated a design curves for optimal design of an anchored sheet pile wall in cohesionless soils based on the results of geotechnical investigation of soil at six different sites using free earth support approach. It was shown that the penetration depth decreased with the increase in angle of internal friction. The tension force in the anchor was also found to increase with the increase in angle of internal friction up to a maximum value of about 250 kN/m which corresponds to an optimum value of angle of internal friction of about 25°. The design curves had showed the efficiency of the optimum design of an anchored sheet pile wall, the values of the design parameter may vary if the free wall height exceeds 7.5 m used in the study. Sable and Patil [7] (2012) developed a program for the analysis and design of low-cost or low-weight cantilever reinforced concrete retaining walls with and without base shear key using Matlab optimtool. The comparison between the conventional design from known designer and optimum cost and weight values which were observed from optimtool showed effectively by cost and weight minimization model and graph. As per conventional design, cost can be reduced by optimization 15% to 30% and weight by 33% to 38% for cantilever retaining wall of height 3 to 6.5 m.

In this research work, a new type of retaining walls is considered. It is comprised of the same three basic elements: stem, toe and heel, where the stem is hinged to the base and tied to the heel by multiple tie rods at intervals along the wall. The tied back retaining wall is suggested to consist of a precast stem of high heights and the base is cast in situ.

OPTIMIZATION SCHEME

Optimization is the act of obtaining the best result under given circumstances [1] . There are many methods that are used to solve the optimization problems [8] . In recent years, some optimization methods that are conceptually different from the traditional mathematical programming techniques have been developed. These methods are labeled as modern or nontraditional methods of optimization. Most of these methods are based on certain characteristics and behavior of biological, molecular, swarm of insects, and neurobiological systems.

One of these modern methods is the genetic algorithm (GA) method (John Holland, 1975) which has been utilized to analyze tied back retaining walls. Genetic algorithms are stochastic search methods that mimic some of the processes of natural biological evolution [9] . The primary purpose of using genetic algorithms is optimization. Genetic algorithms operate on a population of potential solutions applying the principle of "survival of the fittest" to produce better approximations to a solution. Genetic algorithms model natural processes [9] , such as selection, recombination, and mutation.

The stages of the optimization scheme consist of consideration of structural stability, general stability, geotechnical stability, and costs. The required parameters are given subsequently.

FORMULATION FOR OPTIMAL DESIGN

In optimal design problem of retaining wall the aim is to minimize the construction cost of the wall under constraints. This optimization problem can be expressed as follows:

Minimize f(X) which is subject to

$$g_i(X) \le 0 \quad i = 1, 2, \cdots, p \tag{1}$$

$$h_j(X) = 0 \quad j = 1, 2, \cdots, m \tag{2}$$

$$L_k \le X_k \le U_k \quad k = 1, 2, \cdots, n \tag{3}$$

where X is n-dimensional vector called the design vector, $f(X)$ is the objective function, and $g_i(X)$ and $h_j(X)$ are inequality and equality constraints respectively and L_k, U_k are lower and upper bound constraints. To economic design of retaining wall, the objective function, design variables and design constraints should be defined explicitly.

Input Parameters

For the analysis of wall, the geometry of the tied back wall can generally be described by a set of quantities, some of which are viewed as variables during the optimization process. Some quantities are fixed during the process and called "pre-assigned parameters". These are outlined as:

- The properties of backfill and base soil.
- Height of tied back retaining wall.
- Unit weight of concrete, steel and soil.
- The minimum cover for the reinforcement of stem and base.
- The compressive strength of the concrete and the yield strength of the steel to be used in the design.
- The allowable tensile stress of tie steel.
- The cost of each concrete, reinforcing steel, tie steel, formwork, excavation and backfilling works.
- Strength reduction factor (Ø).

Design Variables

The design variables are divided into two categories: those that prescribe the geometric dimensions of wall cross-section, and those that model the steel reinforcement. In general, there are nine design variables of the geometric dimensions to be optimized as shown in Figure 1: T_s: thickness of stem, T_b: thickness of base, B_t: length of toe, B_h: length of heel, D_v: vertical distance between ties, D_h: horizontal distance between ties, α: ratio of height of top of outer tie to wall height (H_1), β: ratio of horizontal distance of outer tie to length of heel (B_h), S: tie spacing.

Whereas the steel reinforcement design variables are modeled as a set of discrete values and include: A_{t1}, A_{t2}: the area of outer and inner tie section respectively, A_{s1}: the area of vertical reinforcement at the stem for negative bending moment (active face of wall), A_{s2}: the area of vertical reinforcement at the stem for positive bending moment (passive face of wall), A_{s3}: the area of longitudinal reinforcement at the stem in the two faces for temperature and shrinkage, A_{s4}, A_{s5}: the area of longitudinal reinforcement at the outer tie level in the stem at tie point and between ties respectively, A_{s6}, A_{s7}:

the area of longitudinal reinforcement at the inner tie level in the stem at tie point and between ties respectively, A_{s8}: the area of main reinforcement at the bottom of the toe, A_{s9}: the area of main reinforcement for negative bending moment at the heel, A_{s10}: the area of main reinforcement for positive bending moment at the heel, A_{s11}: the area of longitudinal reinforcement at the base in the two faces for temperature and shrinkage, A_{s12}, A_{s13}: the area of longitudinal reinforcement at the outer tie in the heel at tie point and between ties respectively, A_{s14}, A_{s15}: the area of horizontal reinforcement at the inner tie level in the base at tie point and between ties respectively.

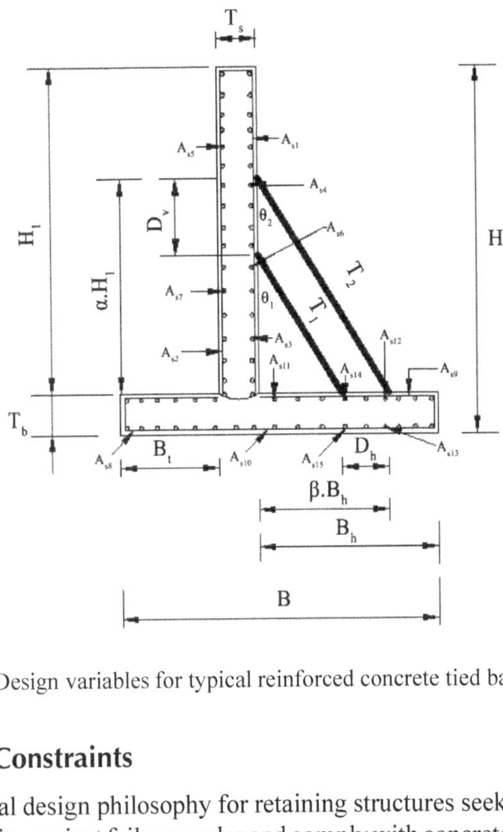

Figure 1: Design variables for typical reinforced concrete tied back retaining wall.

Design Constraints

The typical design philosophy for retaining structures seeks designs that provide safety and stability against failure modes and comply with concrete building code requirements. These requirements may be classified into two groups of constraints namely, the general constraints and the ultimate resistance constraints. Within the optimization procedure, if all considered constraints and cost considerations are entirely met, the design will be feasible. These requirements represent the failure modes as a function of the design variables. Failures modes are summarized in Table 1.

Some constraints have maximum allowable value much larger than that of other constraints. This will badly affect the convergence rate during the minimization of cost function. Therefore, normalization is used which gives after rearranging the equations of these constraints.

Sliding Failure Mode

The net horizontal forces must be such that the wall is prevented from sliding along its foundation. The most significant sliding force component usually comes from the lateral earth pressure acting on the active (backfill) side of the wall. Sliding failure is a result of excessive lateral earth pressures with relation to retaining wall resistance thereby causing the retaining wall system to move away (slide) from the soil it retains.

$$g_1(x) = 1 - \frac{F_r + P_p}{F_s * P_{a2}} \leq 0 \tag{4}$$

where F_r: friction force, P_p: passive force, P_{a2}: driving force, F_s: 1.25 to 2 depending on the importance of the wall, say 1.5 as recommended by Bowles for retaining walls structures [10] .

Overturning Failure Mode

The stabilizing moments, due to vertical forces must be greater than the overturning moments, due to horizontal forces to prevent rotation of the wall around its toe. The stabilizing moments result mainly from the self-weight of the structure, whereas the main source of overturning moments is the active earth pressure. Overturning failure is a result of excessive lateral earth pressures with relation to retaining wall resistance thereby causing the retaining wall system to topple or rotate (overturn).

$$g_2(x) = 1 - \frac{M_c + M_s + M_p}{F_o + M_a} \leq 0 \tag{5}$$

where M_c: total moments of concrete weights, M_s: total moments of soil weights, M_p : total moments of passive force, M_a: active moment, and F_o: 1.5 to 2 depending on the importance of the wall, say 1.5 as recommended by Bowles for retaining walls structures [10] .

Excessive High Toe Pressure

To avoid excessive high toe pressure it is desirable to keep the resultant of the various forces acting on the wall where B: base width, $\overline{X}$: distance of normal resultant force from edge of toe.

Table 1: Failures modes of the problem

Inequality constraints	Failure mode
$g_1(x)$	Sliding stability
$g_2(x)$	Overturning stability
$g_3(x)$, $g_4(x)$	Excessive high toe pressure
$g_5(x)$	Bearing capacity
$g_6(x)$ to $g_{10}(x)$	Flexural moments in stem and base respectively
$g_{11}(x)$ to $g_{23}(x)$	Minimum reinforcement area criteria
$g_{24}(x)$ to $g_{36}(x)$	Tension controlled sections
$g_{37}(x)$, $g_{38}(x)$	Shear failure in stem and base respectively

R within the middle third of the base, so that no tension is developed at the base [10] .

$$g_3(x) = 1 - \frac{2*B}{3*\bar{X}} \leq 0 \tag{6}$$

$$g_4(x) = 1 - \frac{3*\bar{X}}{B} \leq 0 \tag{7}$$

Bearing Capacity

The bearing capacity of the foundation must be large enough to resist the stresses acting along the base of the structure.

$$g_5(x) = 1 - \frac{q_u}{F_b * q_{app}} \leq 0 \tag{8}$$

where q_u : ultimate bearing capacity of the base soil (kN/m^2), q_{app} : the maximum applied pressure (kN/m^2), F_b : safety factor against bearing capacity failure.

Moment Failure Mode

The maximum bending moments at the critical sections should be less than the resistance moments for both stem and base.

$$g_6(x) \text{ to } g_{10}(x) = 1 - \frac{\varnothing M_r i}{M_{app} i} \leq 0 \quad i = 6,7,\cdots,10 \tag{9}$$

where $M_r i$: ultimate bending moments for both stem and base, $M_{app} i$: applied moments at the critical sections of stem and base, $\varnothing$ is the strength reduction factor (assumed to be 0.9) [11] . The section capacity can be calculated from the stress-strain diagrams shown in Figure 2,

$$M_{ri} = A_{si} f_y d_i \left[1 - \frac{A_{si} f_y}{1.7 f'_c b_i d_i} \right] \quad (10)$$

i = index indicating the critical sections of toe, heel and stem,

A_{si}= the area of tensile reinforcement required at the sections,

f_y= yield strength of steel,

f'_c = cylinder compressive strength of concrete,

b_i= distance between ties (S) of the critical sections of toe, heel and stem in the main direction [for the longitudinal direction b_i is supposed to be equal d_i (effective depth of critical sections)],

d_i= effective depth of critical sections [for stem d_i equals to $d_1 = T_s * 1000 - d_b/2 - 50$ and for base d_i equals to $d_2 = T_b * 1000 - d_b/2 - 75$].

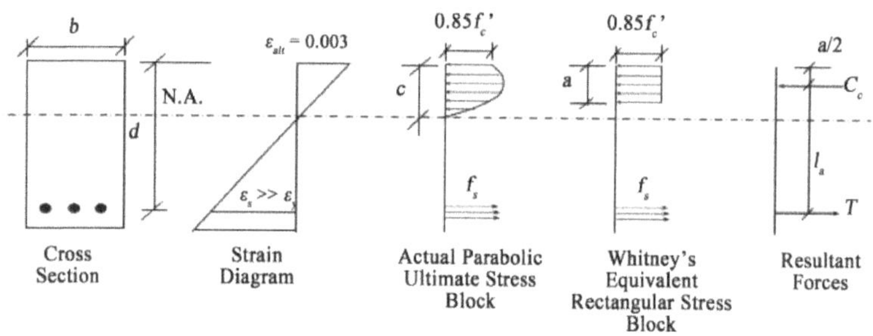

Figure 2: Equivalent rectangular stress block.

Minimum Area of Reinforcement Criteria

The reinforcement area for bending moment at any section where tensile reinforcement is required should not be less than ($A_{s,min}$) [11] .

$$g_{11}(x) \text{ to } g_{23}(x) = 1 - \frac{A_s i}{A_{s,min}} \leq 0 \quad i = 11, 12, \cdots, 23 \quad (11)$$

where ($A_{s,min}$) according to ACI-318-2011 is given by:

$$A_{s,min} = \frac{0.25\sqrt{f'_c}}{f_y} b_i d_i \geq \frac{1.4}{f_y} b_i d_i \quad (12)$$

Tension Controlled Sections

Sections are tension-controlled if the net tensile strain in the extreme tensile steel (ε_t) is equal to or greater than 0.005 when the concrete in compression reaches its assumed strain limit of 0.003 [11] .

$$1-\frac{0.31875*\beta_1*f_c'*b_i*1000*d_i}{A_{s2}*f_y}\leq 0 \qquad i=24,25,\cdots,36 \tag{13}$$

where β_1 equals 0.85.

Shear Failure Mode

The available effective depth must be greater than that required for wide beam shear requirement [11] . So the effective depth for both stem and base must checked as follows,

$$g_{37}(x)=1-\frac{0.17*0.75*d_1*S*\sqrt{f_c'}}{V_s*L_F}\leq 0 \tag{14}$$

$$g_{38}(x)=1-\frac{0.17*0.75*d_2*S*\sqrt{f_c'}}{V_b*L_F}\leq 0 \tag{15}$$

where S: distance between ties, L_F: load factor, V_s : the maximum actual shear at stem for S meters (kN), V_b: the maximum actual shear at base for S meters (kN).

Objective Function

The objective function is a function of design variables the value of which provides the basis for choice between alternate acceptable designs. The objective of design may be minimization of weight, cost or stress concentration factor. In structural designs the objective function is usually weight or cost minimization.

In the present study, the objective function is defined as the total cost of tied back retaining wall (material and labor) for a spacing length (S) of the wall. This includes the cost of concrete (C_{conc}), cost of various steel reinforcement, cost of tie steel (C_{ti}), cost of formwork (C_{fw}), cost of excavation works (C_{ex}), and cost of backfilling works (C_{fi}).

$$f(x)=V_c\cdot P_c+V_{st}\cdot\rho_s\cdot P_s+V_{ti}\cdot\rho_s\cdot P_s+A_{fw}\cdot P_{fw}+V_{ex}\cdot P_{ex}+V_{fi}\cdot P_{fi} \tag{16}$$

where V_c: volume of concrete, P_c : price of concrete per cubic meter, V_{st}: volume of reinforcing steel, ρ_s = unit mass of steel (ton/m^3), P_s = price of steel per ton, V_{ti}: volume of tie steel, A_{fw}: area of concrete that used for form working, P_{fw} : price of formwork per square meter, V_{ex}: volume of excavation, P_{ex}= price of excavation per cubic meter, V_{fi}: volume of backfilling, P_{fi}= price of backfilling per cubic meter.

APPLICATION OF DEVELOPED PROGRAM

The applications in this section involve solving many numerical examples in order to illustrate the effects of various design variables and different parameters on the optimal design. The minimum cost of the tied back retaining wall for a distance equal to the spacing between ties (S) is given. These examples are concerned with the following points:

- The effect of the total height of the tied back retaining wall.
- The effect of pressure coefficient, K, which is a function of backfill properties, i.e. $(K = \gamma_f \cdot k_a)$.
- The effect of materials properties (concrete, reinforcing steel, and tie steel).

The program consists of two main stages. In the first stage the necessary data are specified. The second stage performs the calculations of the optimum design.

Effect of the Total Height of Wall

To investigate the effect of the total height of the tied back retaining wall, height equal to (6, 7, 8, 9 and 10) m is used. The basic values of required parameters are taken as pressure coefficient K = 6 kN/m^3, yield stress of steel f_y = 276 MPa, concrete cylinder compressive strength $f'_c = 25$ MPa, allowable tensile stress of tie steel σ_{all} = 155 MPa, unit mass of steel ρ_s = 7.85 ton/m^3, unit weight of reinforced concrete γ_c = 24 kN/m^3, unit weight of base soil γ_b = 19 kN/m^3, cohesion of base soil c_b = 40 kN/m^2, angle of internal friction for base soil = 35°, cost of steel P_s = 1,000,000 ID/ton, cost of concrete P_c = 150,000 ID/m^3, cost of formwork P_{fw} = 7500 ID/m^2, cost of excavation P_{ex} = 6000 ID/m^3, cost of backfilling P_{fi} = 10,000 ID/m^3.

The optimal solution has been summarized in Table 2 and Figure 3 in which the height represents the total height of tied back retaining wall H_2.

It can be noticed from this table and figure that the total cost of wall increases as the total height H_2 increases. The data from this table and figure also shows nonlinear relationship between the total cost of wall and the total height H_2. This relation usually depends on many factors like material properties and the different unit prices as will be discussed in following sections. The total height of tied back retaining wall has affected on the optimum stem thickness T_s which increases with the increase of total height H_2. Also the thickness of the base T_b increases as the wall height increases.

The Effect of Pressure Coefficient

In order to examine the effect of backfill properties (γ_f, φ_f) on the optimum cost and other optimum values, different values of pressure coefficient which is a function of these backfill properties are considered. Pressure coefficients values (in addition to 6 kN/m^3 that was used in Section 4.1) equal to 4, 4.5, 5, and 5.5 kN/m^3 are used in order to illustrate the effect of its variation on the optimum results with different total height that were considered in previous section. The other data remains unchanged as in the previous section. The results have been shown in Figures 4-10.

It is clear from Figure 4 that the minimum cost increases according to the increase of wall height. The influence of pressure coefficient increases as the wall becomes higher. From Figure 5 it is clear that the tie spacing is inversely proportional to wall height and it can be noticed that the effect of pressure coefficient vanishes as the wall becomes higher. It is clear from Figure 6 that the stem thickness increases according to the increase of wall height. The increasing of pressure coefficient leads to increase the stem thickness in same rate as the wall becomes higher. From Figure 7, it is clear that the base thickness increases with the increase of wall height. Also it can be noticed, in general, that increasing pressure coefficient seems to have little effect on base thickness up to 5 kN/m^3 and the effect appears after that. It can be seen that the effect of height vanishes for wall heights 9 m and 10 m where the curves coincide. It can be realized from Figure 8 that the base width is proportional to wall height. Also there is no effect of pressure coefficient up to 5 kN/m^3 and after that the influence of pressure coefficient is increasing obviously. It is clear from Figure 9 that the outer tie position ratio from H_1, α, is inversely proportional to wall height and it seems not to alter as the pressure coefficient varies. It is clear from Figure 10 that the increase in wall height and pressure coefficient has a very little effect on the outer tie position ratio from B_h, β, it seems to be constant, about (0.92 to 1) times B_h.

The Effect of Materials Properties

The properties of concrete, reinforcing steel, and tie steel play an important role on the optimum design. Their effects have arisen not only in unit price of each material but also in their specifications that influence the optimal design as will be shown in this section.

In order to examine the effect of concrete compressive strength on the optimum results different grades of concrete are considered. Grades (25, 30, 40, and 50 MPa), which indicate the cylinder compressive strength, are investigated.

Table 2: The optimum design dimensions and various steal areas for various total height (6 to 10) m of tied-back retaining wall

H_t	Cost*10^6	T_s	T_b	B_t	B_h	D_v	D_h	α	β	S	A_{t1}	A_{t2}	A_{s1}	A_{s2}	A_{s3}	A_{s4}	A_{s5}	A_{s6}	A_{s7}	A_{s8}	A_{s9}	A_{s10}	A_{s11}	A_{s12}	A_{s13}	A_{s14}	A_{s15}
m	I.D/S	m	m	m	m	m	m			m	mm^2	mm^2	mm^2	mm^2	mm^2	mm^2	mm^2	mm^2	mm^2	mm^2	mm^2	mm^2	mm^2	mm^2	mm^2	mm^2	mm^2
6	1.513	0.26	0.328	1.496	3.084	2.3	1.237	0.756	0.961	1.541	984	372	2060	4424	3750	1375	1380	1376	1374	6830	2062	4076	2955	1100	1197	1115	1125
7	1.823	0.29	0.423	2.729	2.291	2.6	0.588	0.679	0.926	1.53	1129	412	3417	4476	1752	1515	1121	1126	1156	1185	8800	2585	2590	987	1450	995	992
8	2.116	0.311	0.456	1.978	2.638	2.721	0.551	0.634	0.974	1.5	1138	436	1862	2540	4923	2540	1710	1789	1720	10.410	2540	2644	1117	1292	1292	1745	2540
9	2.92	0.326	0.514	1.826	3.443	0.522	0.517	0.611	0.989	1.45	1178	457	2076	5198	5652	1970	1982	1975	1995	12.651	2890	7405	5182	1659	1712	1665	1779
10	3.264	0.334	0.674	2.52	3.909	1.979	0.534	0.56	0.961	1.411	1191	471	2151	1980	5275	1129	1145	1130	1155	16.919	1688	9545	7685	2484	2565	2489	2578

The remaining parameters are considered to have the same values of Section 4.1. Since practically the increase in grade of concrete leads to increase its cost, so the effect of unit cost of concrete is combined with the influence of concrete compressive strength here. The unit costs of concrete considered in this study for the specified compressive strengths (25, 30, 40, and 50 MPa) are (150,000, 250,000, 500,000, 700,000) I.D/m^3 respectively.

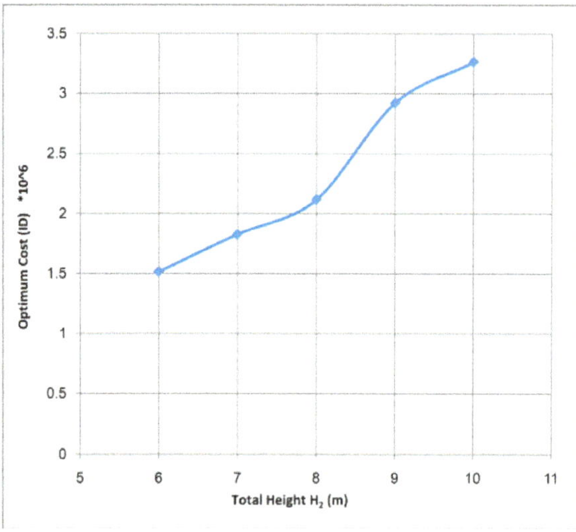

Figure 3: Optimum cost versus total wall height, H_2.

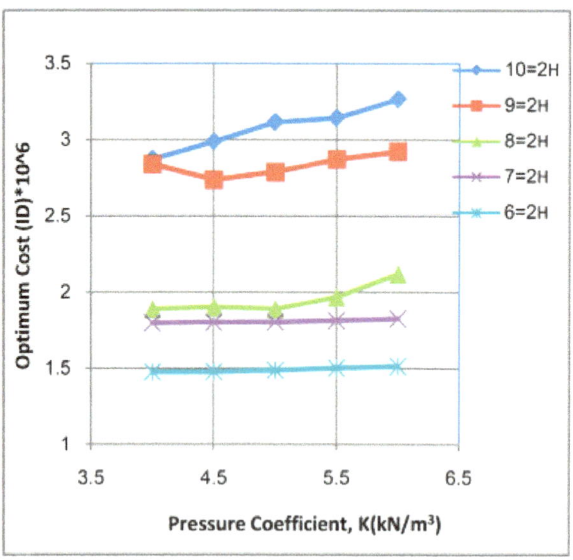

Figure 4: Optimum cost versus pressure coefficient, K, for different values of wall height, H_2.

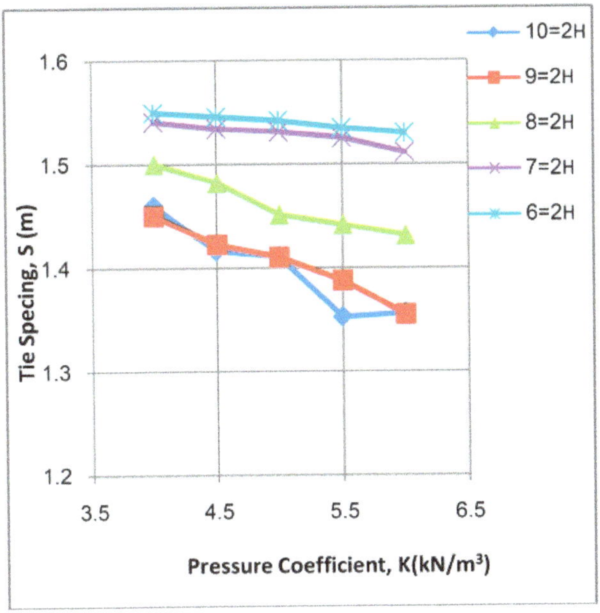

Figure 5: Tie spacing, S, versus pressure coefficient, K, for different values of wall height, H_2.

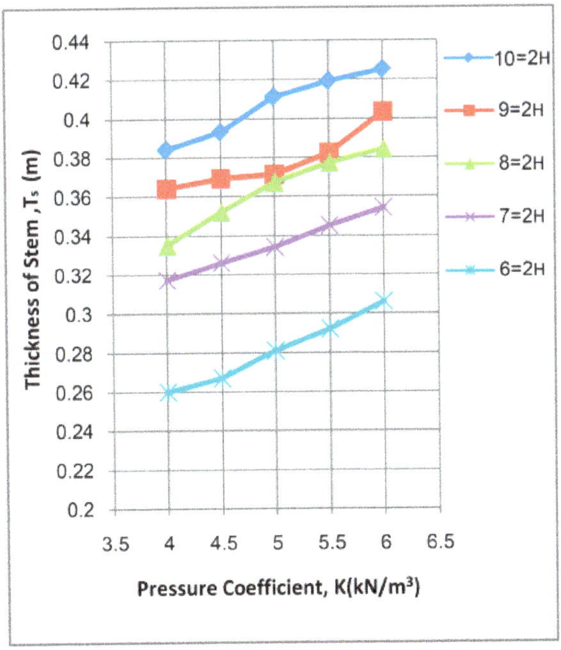

Figure 6: Stem thickness, T_s, versus pressure coefficient, K, for different values of wall height, H_2.

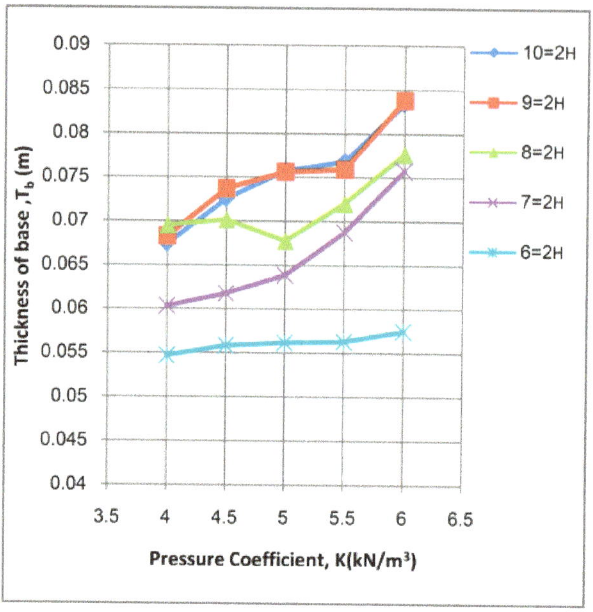

Figure 7: Base thickness, T_b, versus pressure coefficient, K, for different values of wall height, H_2.

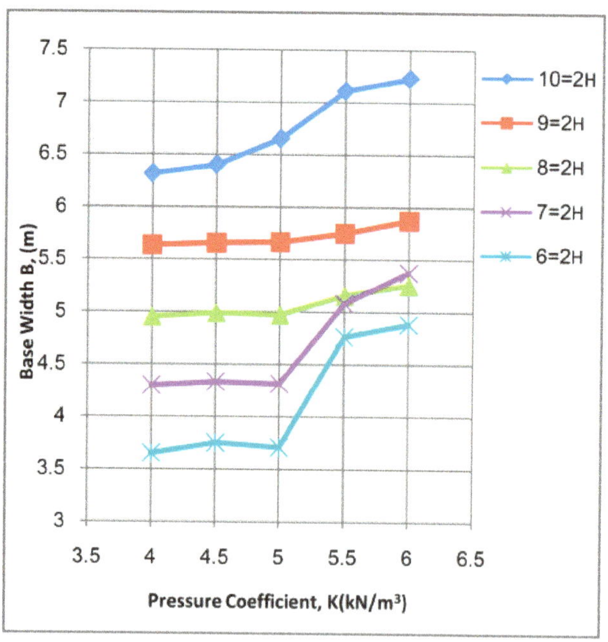

Figure 8: Base width, B, versus pressure coefficient, K, for different values of wall height, H_2.

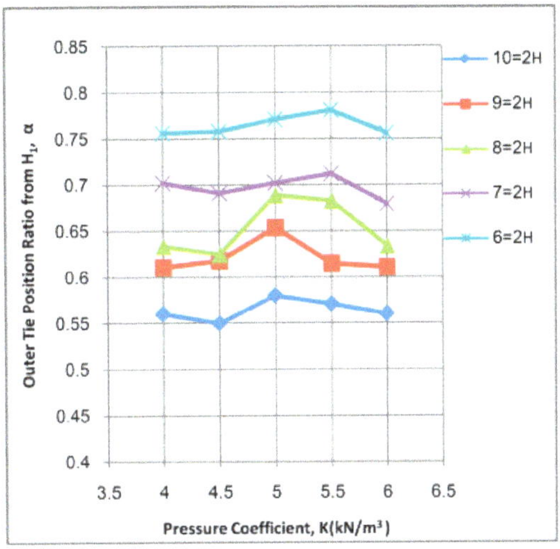

Figure 9: Outer tie position ratio from H_1, α, versus pressure coefficient, K, for different values of wall height, H_2.

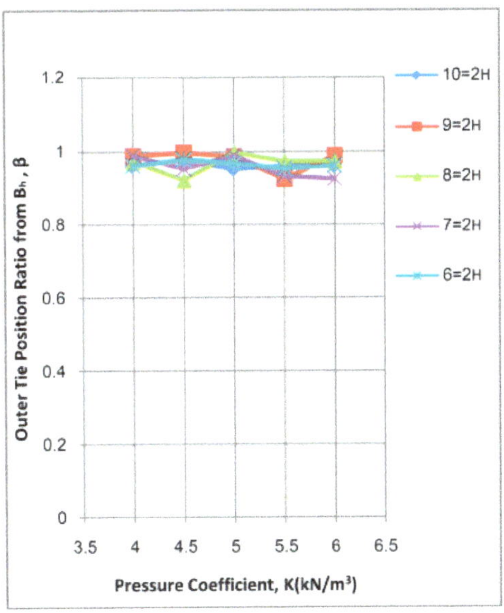

Figure 10: Outer tie position ratio from B_h, β, versus pressure coefficient, K, for different values of wall height, H_2.

For steel reinforcement yield strengths of (276, 350, 415, and 460 MPa) are used, and also the same information data of Section 4.1 are utilized.

Considering the allowable stress of tie steel, the values of allowable stresses are given by BS449: Part 2: 1969 [12] . The values of 140, 215 in addition to 155 MPa that is used in Section 4.1 are chosen in order to illustrate the effect of its variation on the optimum dimensions. The results obtained by varying the properties of materials are shown in Figures 11-13.

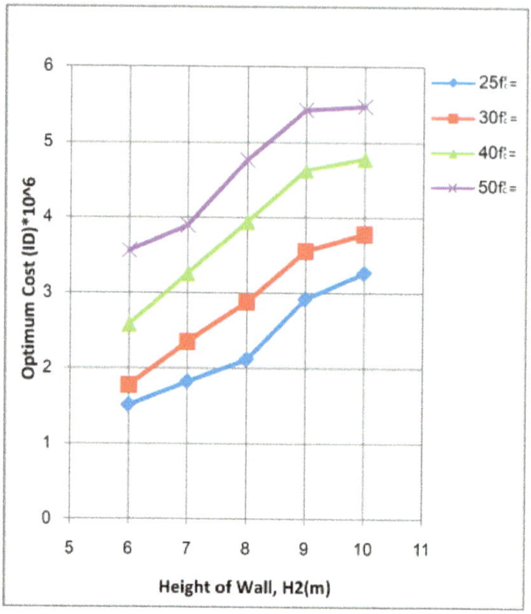

Figure 11: Optimum cost versus total height, H_2, for different compressive strength values, f'_c.

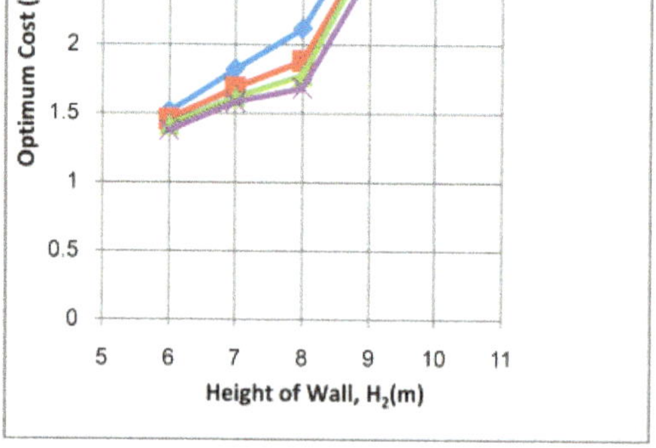

Figure 12: Optimum cost versus total height, H_2, for different yield strength values, f_y.

It can be realized that the compressive strength of concrete has an effect on the optimum thickness of stem and base. Increasing the concrete compressive strength leads to a reduction in the stem thickness and a reduction in the base thickness, consequently different steel areas are changed in each section.

Hence, it is clear from Figure 11 that the increasing of the concrete compressive strength leads to increase the minimum cost. Results reveal that the increase of steel strength leads to an effect on the dimensions and steel areas in the optimum sections, and the increase of the yield steel strength leads to a decrease in the minimum cost as notice from Figure 12. Therefore, the results indicate that it is economical to use steel of high strength in design.

It is clear from Figure 13 that the increase of the allowable stress of tie steel leads to decrease the minimum cost. Therefore, it is economical to use tie steel section with high allowable stress in design.

PROPORTIONS OF TIED BACK RETAINING WALL

Dimensions of tied back retaining wall should be adequate for structural stability and to satisfy design requirements. So there is a need for tentative dimensions that may be used by the designer in the absence of other data though overly conservative.

Based on the used pressure coefficient values (4 to 6) kN/m^3 and the wall total height (6 to 10) m, the developed program is utilized to find the values or ranges of some parameters relating to tied back retaining wall which give an optimum of such retaining structures. Tables 3-8 show some the results of this study.

CONCLUSIONS

The following conclusions may be drawn from the present study:

1) The minimum cost of tied back retaining wall increases as the total height H_2 increases in a nonlinear relationship. This relation usually depends on many factors like material properties and the different unit prices.
2) The ranges of optimum dimensions as ratios from the wall height H_2 for a tied back retaining wall are (0.135 to 0.257) for the distance between ties S, (0.038 to 0.051) for the thickness of the wall T_s, (0.055 to 0.083) for the thickness of the base T_b, and (0.609 to 0.794) for the width of the base B.

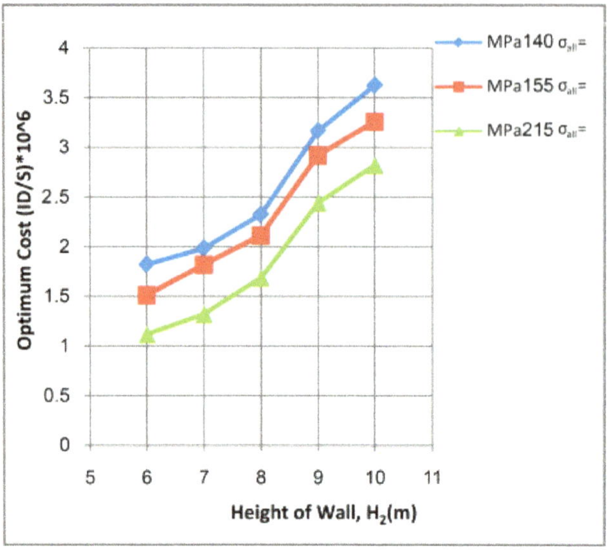

Figure 13: Optimum cost versus total height, H_2, for different allowable stress values, σ_{all}.

Table 3: The optimum values of tie spacing to wall height ratio, S/H_2

	S/H_2 ratios				
K \ ***H₂***	6	7	8	9	10
4	0.257	0.221	0.188	0.161	0.146
4.5	0.256	0.221	0.185	0.158	0.142
5	0.255	0.22	0.181	0.157	0.141
5.5	0.254	0.219	0.18	0.154	0.135
6	0.252	0.219	0.179	0.15	0.136

Table 4: The optimum values of stem thickness to wall height ratio, T_s/H_2

	T_s/H_2 ratios				
K \ ***H₂***	6	7	8	9	10
4	0.043	0.045	0.042	0.04	0.038
4.5	0.045	0.047	0.044	0.041	0.039
5	0.047	0.048	0.046	0.041	0.041
5.5	0.049	0.049	0.047	0.042	0.042
6	0.051	0.051	0.048	0.045	0.043

Table 5: The optimum values of base thickness to wall height ratio, T_b/H_2

T_b/H_2 ratios					
K \ H_2	6	7	8	9	10
4	0.055	0.06	0.07	0.068	0.067
4.5	0.056	0.062	0.07	0.074	0.073
5	0.056	0.064	0.068	0.076	0.076
5.5	0.056	0.069	0.072	0.076	0.077
6	0.058	0.076	0.078	0.084	0.083

Table 6: The optimum values of base width to wall height ratio, B/H_2

B/H_2 ratios					
K \ H_2	6	7	8	9	10
4	0.609	0.614	0.619	0.626	0.631
4.5	0.625	0.619	0.624	0.628	0.64
5	0.618	0.616	0.621	0.629	0.665
5.5	0.794	0.726	0.644	0.639	0.711
6	0.814	0.768	0.657	0.652	0.723

Table 7: The optimum values of upper tie position ratio, α

α values					
K \ H_2	6	7	8	9	10
4	0.756	0.702	0.634	0.611	0.56
4.5	0.758	0.691	0.625	0.55	0.55
5	0.771	0.702	0.689	0.654	0.58
5.5	0.781	0.712	0.683	0.615	0.571
6	0.756	0.679	0.634	0.611	0.56

Table 8: The optimum values of upper tie position ratio, β

β values					
K \ H_2	6	7	8	9	10
4	0.961	0.989	0.974	0.989	0.961
4.5	0.978	0.954	0.923	0.998	0.975
5	0.968	0.989	0.998	0.987	0.952
5.5	0.954	0.933	0.974	0.926	0.961
6	0.961	0.926	0.974	0.989	0.961

3) The outer tie position ratio, α, is from (0.56 to 0.781) times H_1 and the outer tie position ratio, β, seems to be constant, about (0.92 to 1) times B_h.

4) It can be realized from Figures 4-13 that, there is a real irregularity in the trends of the relationships. This can attribute to the nature of the objective function, since the minimization of the cost is the main task of the program. The big

number of design variables and selecting the optimum results from different methods are other minor reasons.

5) From cost point of view it is preferable to use low strength concrete and steel of high yield strength in the design of tied back retaining wall.

6) The increase of the allowable stress of tie steel leads to the decrease of the minimum cost. Therefore, it is economical to use tie steel section with high allowable stress in design.

REFERENCES

1. Rao, S.S. (2009) Engineering Optimization: Theory and Practice. 4th Edition, John Wiley & Sons, Inc., Hoboken, New Jersey. http://dx.doi.org/10.1002/9780470549124
2. Al-Janabi, A.I. (1983) The Structural Analysis of Tied Back Retaining Walls by Finite Element Method. M.Sc. Thesis, Universuty of Basrah, Basrah, Iraq.
3. Ceranic, B. and Fryer, C. (1999) A Constrained Simulated Annealing Approach to the Minimum Cost Design of Reinforced Concrete Cantilever Retaining Walls. Proceedings of 3rd World Congress on Structural and Multi-Disciplinary Optimization, New York, 593-595.
4. Ghazavi, M. and Heidarpour, A. (2003) Optimization of Counterfort Retaining Walls. Fourth International Conference of Earthquake Engineering and Seismology, Tehran, 2003.
5. Yepes, V., Alcala, J., Perea, C. and Vidosa, F. (2007) A Parametric Study of Optimum Earth-Retaining Walls by Simulated Annealing. Engineering Structures, 30, 821-830. http://dx.doi.org/10.1016/j.engstruct.2007.05.023
6. Magbo, C.C., Sule, S. and Nwofor, T.C. (2012) Design Curves for Optimal Design of Anchored Sheet Pile Walls in Cohessionless Soils. Canadian Journal on Environmental, Construction and Civil Engineering, 3, 104-112.
7. Sable, K.S. and Patil, A. (2012) Comparision between Optimization and Conventional Catilever Retaining Wall by Using Optimtool in Matlab. International Journal of Engineering and Advanced Technology (IJEAT), 1, 255.
8. Goldberg, D.E. (1989) Genetic Algorithms in Search, Optimization, and Machine Learning. Addison-Wesley Publishing Company, USA.
9. Sivanandam, S.N. and Deepa, S.N. (2008) Introduction to Genetic Algorithms. Verlag, Berlin Heidelberg.
10. Bowles, J.E. (1986) Analytical and Computer Methods in Foundation Engineering. MsGraw-Hill, New York.
11. American Concrete Institute (2011) Building Code Requirements for Structural Concrete (ACI 318M-11) and Commentary, ACI Committee 318.
12. British Standards Institution (1969) The Use of Structural Steel in Building. BS 449, Part 2.

2

CHAPTER

BEHAVIOR OF REINFORCED RETAINING WALLS WITH DIFFERENT REINFORCEMENT SPACING DURING VEHICLE COLLISIONS

Kwangkuk Ahn and Hongsig Kang

INTRODUCTION

Continuously increasing volumes of traffic have resulted in problems of traffic congestion. To solve this problem, a large number of underpasses, crossroads, and elevated roads have recently been constructed. Small and large reinforced retaining walls, which are both economical and beneficial in terms of construction efficiency, are being constructed along the sides of roads. As the construction of reinforced retaining walls along roads that cater to high traffic volume increases, vehicle collisions with these walls also increase.

A reinforced retaining wall is constructed by gradually increasing the tensile force applied to the soil on the backside of the wall. In this method, the wall is reinforced by using high-tensile-strength reinforcing materials such as geotextiles. The material type, construction interval, and length of the reinforcing material are considered important design factors when designing a reinforced retaining wall. However, a reinforced wall is built of blocks and is not one united body, so a vehicle colliding with it can potentially cause secondary damage owing to the collapse of the retaining wall and the settlement of backfill.

In order to investigate the characteristics of behavior and failure mechanisms of reinforced retaining walls, model tests and numerical analyses were carried out by changing the length and spacing of reinforcement, surcharge load, and backfill properties studied by Wong, Pinto, Leshchinsky, Ghionna, Yoo, Liu, Stuedlein, and Suliman [1–8]. The safety of construction structures according to vehicle collision was verified with vehicle crash tests and numerical collision analyses by Wu, Borovinšek, Chung, Itoh, Tay, and Kim [9–14]. However, the effect of vehicle collision on reinforced retaining

Citation: Kwangkuk Ahn and Hongsig Kang, "Behavior of Reinforced Retaining Walls with Different Reinforcement Spacing during Vehicle Collisions," Advances in Materials Science and Engineering, vol. 2015, Article ID 920628, 9 pages, 2015. doi:10.1155/2015/920628.

walls was excluded. Additionally, the safety of reinforced retaining walls affected by vehicle collision for different vehicle velocities was recently studied by Ahn et al. [15]; however, the effect of design factors on reinforced retaining walls was not considered.

In this study, a collision analysis was conducted to identify the behavior of reinforced retaining walls with spacing of reinforcement when a vehicle crashed into the reinforced retaining wall. This analysis was conducted with three-dimensional nonlinear dynamic time history analysis using LS-DYNA, a general finite-element analysis program, to identify the behavior of reinforced retaining walls during vehicle collision. An eight-ton single-unit Ford truck offered by the National Crash Analysis Center (NCAC) was used for the vehicle collision analysis. Behavior of reinforced retaining walls from vehicle collisions was analyzed by comparing the displacement of the front side, horizontal and vertical, of the reinforced retaining wall's block at the point of collision from the ground surface to 1.0H which is the height of the reinforced retaining wall at intervals of 0.2H.

CONDITIONS OF VEHICLE COLLISION

The guidelines for vehicle collisions found in "guidelines for installation and administration of road safety facility" along with "business guide for vehicle crash test of safety barrier provided in Korea" were considered in this study [16, 17].

Calculation of Impact Severity

The impact severity was defined by the kinetic energy which is created when a vehicle crashes into a structure. Figure 1 shows the condition of vehicle collision and the impact severity is computed as follows:

$$\mathrm{IS} = \frac{1}{2}m\left(\frac{v}{3.6}\sin\theta\right), \qquad (1)$$

where IS is impact severity (kg), m is weight of vehicle collision (ton), v is velocity of vehicle collision (km/h), and θ is angle of vehicle collision (°) [16, 17].

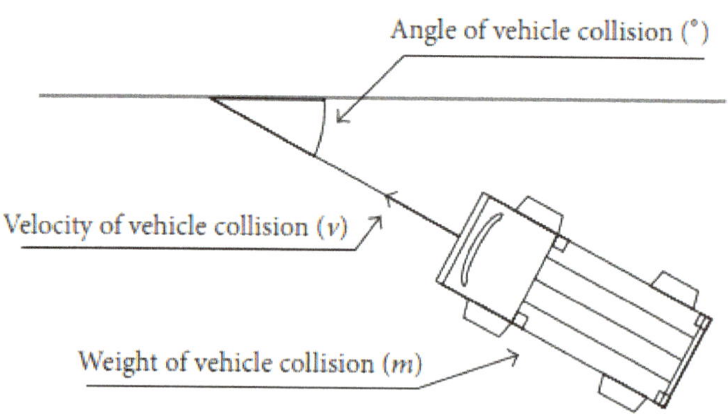

Figure 1: Condition of impact severity.

The Velocity of Vehicle Collision

According to "guidelines for installation and administration of road safety facility in Korea" the vehicle collision velocity can be as high as 80% of the design velocity. Table 1 shows the relationship between the design and collision velocities on the road [17], for each type of vehicle. Therefore 80 km/h of collision velocity was chosen to consider the conditions of the roads in Korea in this study.

Table 1: Relationship between design and collision velocities

Type	The range of design velocity in road (km/h)	Used design velocity (km/h)	Collision velocity (km/h)
A	60~120	100	60
B	60	60	40
C	20~50	50	35
D	—	100	80

Weight of Vehicle Collision

For small vehicles and large vehicles, the "guidelines for installation and administration of road safety facility" laid down rules governing the weights of allowed vehicles in Korea. The weights of the vehicles ranged from 3.5 to 8.0 tons, and these were used as the design weights for the crashed vehicles [16, 17]. Therefore the weight of the collided vehicle was chosen to be 8 tons and the eight-ton Ford single-unit truck, which is the finite-element model offered by the National Crash Analysis Center (NCAC), was used for the vehicle collision analysis in this study.

Angle of Vehicle Collision

The angle which is the included angle between the structure and the crashed vehicle is widely known to be 15° for a straight road and 16° for a curved road. "Guidelines for installation and administration of road safety facility" used 15° for large vehicles and 20° for small vehicles in Korea [17]. Therefore 15° was used for the angle in this study.

NUMERICAL ANALYSIS FOR DIFFERENT REINFORCEMENT SPACING

Numerical Model of Vehicle

In this study, the numerical model used to model the vehicle is an eight-ton truck model (a Ford single-unit truck). It can be modeled using LS-DYNA and is recommended by the NCAC. Figure 2 shows a three-dimensional model of the vehicle. The model consists of 19,479 shells, 124 beams, and 1,248 solid elements. The steel material used has a 270 MPa yield stress and 205 GPa modulus of elasticity, as shown in Table 2 [15].

Table 2: Properties of truck model

Type	Truck
The number of elements	
Shell	19,479
Solid	1,248
Beam	124
Total	20,727
Weight (kg)	8,035
Yield stress (MPa)	270
Modulus of elasticity (MPa)	205,000

Figure 2: Three-dimensional model of Ford single-unit truck.

Properties of the Reinforced Retaining Wall

The model of the reinforced retaining wall which was 20,000 (W) × 10,000 (L) × 5,200 (H) mm consisted of block, reinforcement, crushed stone, and backfill (Figure 3). The block, 400 (W) × 400 (L) × 200 (H) mm, was modeled with solid elements and its material was Elastic_Tiltel in LS-DYNA as in Table 3. Also the contact condition between the blocks is defined by the automatic_single_surface function in LS-DYNA. Reinforcement was modeled with beam elements and its material was Picewise_linear_plasicity in LS-DYNA as in Table 4. The coincident function in LS-DYNA was used for reinforcement to couple the block and backfill. Crushed stone and backfill of the reinforced retaining wall were modeled with solid elements and this material was FHWA_Soil_Title as seen in Table 5. Nodes of each element of crushed stone and backfill were merged to integrate movement. Vehicle collisions were simulated with each model, using a contact algorithm to model the wall behavior. Tables 3, 4, and 5 show details of each created model [15].

Table 3: Properties of reinforcement

Type	Material of model (in LS-DYNA)	Unit weight (kN/m^3)	Modulus of elasticity (MPa)	Poisson ratio
Beam	Piecewise_linear_plasticity	0.1	3,000	0.25

Table 4: Properties of reinforced retaining wall's block

Type	Size (mm)	Material of model (in LS-DYNA)	Unit weight (kN/m^3)	Modulus of elasticity (MPa)	Poisson ratio
Solid	400 × 400 × 200	Elastic_Title	23.5	5000	0.3

Table 5: Properties of crushed stone and backfill

Type	Material of model (in LS-DYNA)	Unit weight (kN/m^3)	Internal friction angle (°)	Bulk modulus of elasticity (MPa)	Shear modulus (MPa)
Crushed stone	FHWA_Soil_Title	19.0	45°	0.465	0.186
Backfill	FHWA_Soil_Title	19.0	30°	0.35	0.14

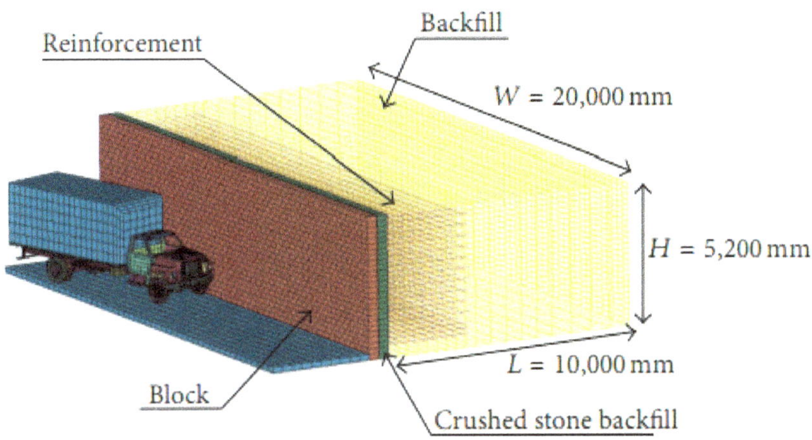

Figure 3: Collision analysis model of reinforced retaining wall.

Spacing of Reinforcement

In general, in the design of reinforced retaining walls, reinforcement is installed and its spacing is denser from the top to the bottom in consideration of the earth pressure. However, as in Figure 4, the reinforcement was placed at equal-intervals of 200, 400, and 600 mm from the top of the reinforced retaining wall to determine the effect of varying intervals of reinforcement on vehicle collision and a collision analysis was conducted.

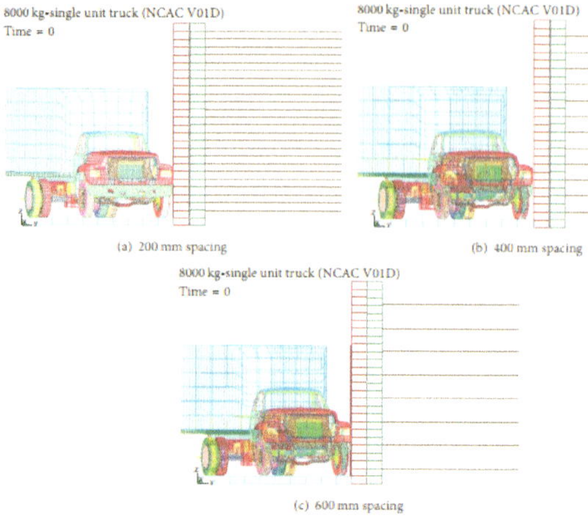

Figure 4: Analysis model for different reinforcement spacing.

The observation point was measured from the ground surface to 1.0 at intervals of 0.2. The reinforcement was not placed at the top of the reinforced retaining wall because it could not perform the function of reinforcement.

RESULTS OF ANALYSIS

Behavior of Reinforced Retaining Wall with Different Reinforcement Spacing

A collision analysis was conducted using LS-DYNA, a general finite-element analysis program, to identify the behavior of the reinforced retaining wall during vehicle collision. The contact condition between the reinforced retaining wall and the truck is defined by the contact_Automatic_surface_to_surface function in LS-DYNA, and the result of the collision simulation is shown in Figure 5.

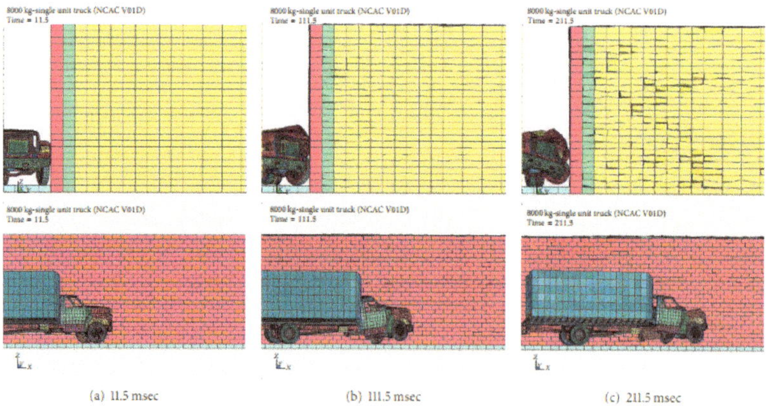

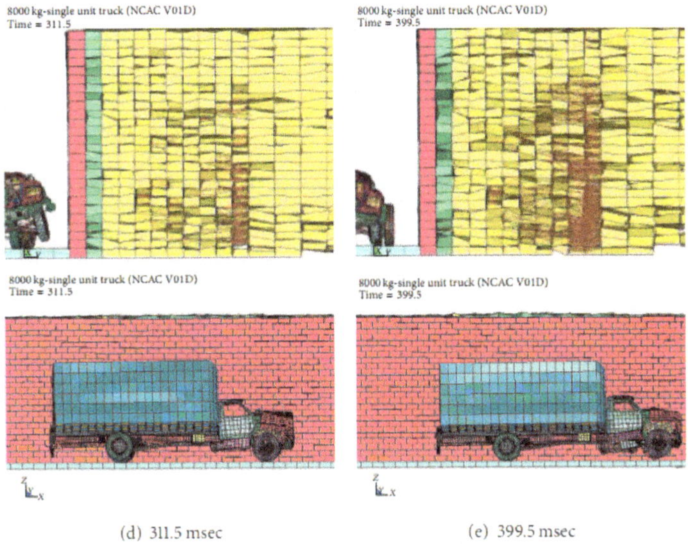

Figure 5: Behavior of reinforced retaining wall during vehicle collision.

Displacement of Block with 200 mm Reinforcement Spacing

Table 6 shows the vertical and horizontal displacement with respect to the height (H) of the reinforced retaining wall block for 200 mm reinforcement spacing and presents the separation ratio by contrasting the original block size with the maximum displacement of the block. Figure 6 shows the vertical and horizontal displacements of the front block, plotted at 0.2H, 0.4H, 0.6H, 0.8H, and H from the bottom of the reinforced retaining wall for 200 mm reinforcement spacing. In Figure 6(a), a positive value of horizontal displacement implies that the block moved toward the backfill of the reinforced retaining wall, and a negative value of horizontal displacement implies that the block moved toward the collided vehicle. In Figure 6(b), a positive value for vertical displacement indicates the bulging of the block and a negative value of vertical displacement indicates the subsidence of the block.

Table 6: Block displacement for 200 mm reinforcement spacing

	Horizontal displacement (mm)			Vertical displacement (mm)		
Height	Maximum horizontal displacement	Separation ratio (%)	Final horizontal displacement	Maximum vertical displacement	Separation ratio (%)	Final vertical displacement
0.2*H*	20.5	5.1	2.6	22.2	11.1	−11.5
0.4*H*	26.6	6.7	26.6	−42.2	21.1	−42.2
0.6*H*	42.8	10.7	42.8	−48.6	24.3	−48.6
0.8*H*	33.2	8.3	30.8	−57.9	29.0	−57.9
H	−41.3	10.3	5.9	−68.1	34.0	−68.1

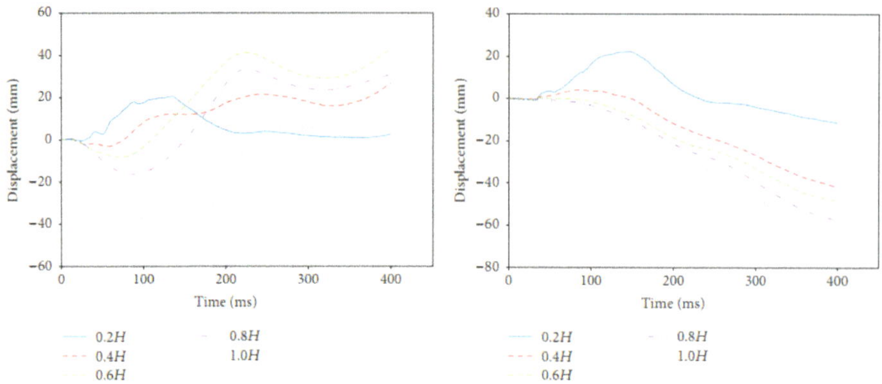

(a) Horizontal displacement of block with respect to elapsed time

(b) Vertical displacement of block with respect to elapsed time

Figure 6: Block displacement for 200 mm reinforcement spacing.

The final horizontal displacement at 0.2H (which was closest to the collision contact surface) was 2.6 mm, and the final horizontal displacement at H (furthest from the collision contact surface) was 5.9 mm. Relatively smaller final horizontal displacements were observed in these two positions on the reinforced retaining wall, as compared to the other positions. The final horizontal displacement at 0.6H was found to be the largest, at 42.8 mm. For horizontal displacement, the separation ratio, which contrasts the original block size with the maximum displacement of the block, ranged between 5.1 and 10.3%. The final vertical displacement was found to be negative, and the displacement increased with height. The final vertical displacement was maximum at H, at –68.1 mm. The separation ratio, in terms of vertical displacement, ranged between 11.1 and 34.0%.

Displacement of Block for 400 mm Reinforcement Spacing

Figure 7 and Table 7 show the vertical and horizontal displacement with respect to the height of the reinforced retaining wall block, for 400 mm reinforcement spacing.

Table 7: Separation ratio for 400 mm reinforcement spacing

Height	Horizontal displacement (mm)			Vertical displacement (mm)		
	Maximum horizontal displacement	Separation ratio (%)	Final horizontal displacement	Maximum vertical displacement	Separation ratio (%)	Final vertical displacement
0.2*H*	18.0	4.5	18.0	−16.1	15.0	−16.1
0.4*H*	8.7	2.2	2.7	−45.3	43.1	−45.3
0.6*H*	36.2	9.1	36.2	−65.3	32.7	−65.3
0.8*H*	51.8	12.9	51.8	−67.6	33.8	−67.6
H	−37.4	9.4	26.5	−69.1	34.6	−69.1

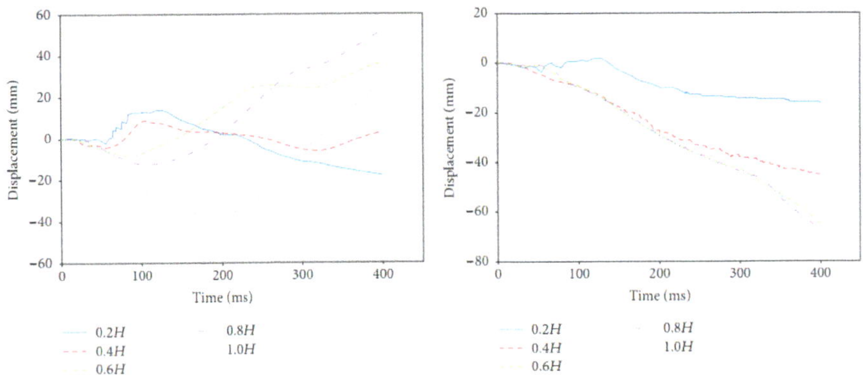

(a) Horizontal displacement of block with respect to elapsed time (b) Vertical displacement of block with respect to elapsed time

Figure 7: Block displacement for 400 mm reinforcement spacing.

The separation ratio, contrasting the original block size with the maximum displacement of the block, in terms of horizontal displacement, ranged between 4.5 and 12.9%. The final horizontal displacement was maximum at 0.8H, at 51.8 mm. The final vertical displacement was again negative throughout and increased in value with the height of the reinforced retaining wall, as was the case for the 200 mm reinforcement spacing. The biggest final vertical displacement observed was –69.1 mm at the top (H) of the reinforced retaining wall. The separation ratio, contrasting the original block size with the maximum displacement of the block, in terms of vertical displacement, ranged between 15.0 and 43.1%. As shown in Figure 7(a), it was possible to observe a restoration to the original state at the heights of 0.2H and 0.4H, after an initial movement into the backfill of the reinforced retaining wall. However, when observing the upper part of the reinforced retaining wall after the collision, it was found that the increase in block displacement had not absorbed the impulsive load of the vehicle due to a malfunction of the backfill and reinforcement.

Displacement of Block for 600 mm Reinforcement Spacing

Figure 8 and Table 8 show the vertical and horizontal displacement as a function of the height of the reinforced retaining wall block, for 600 mm reinforcement spacing.

Table 8: Block displacement for 600 mm reinforcement spacing

	Horizontal displacement (mm)			Vertical displacement (mm)		
Height	Maximum horizontal displacement	Separation ratio (%)	Final horizontal displacement	Maximum vertical displacement	Separation ratio (%)	Final vertical displacement
0.2*H*	−27.0	6.8	−27.0	−20.9	5.2	−20.8
0.4*H*	−39.5	9.9	−39.5	−49.8	12.4	−49.8
0.6*H*	−11.6	2.9	−1.9	−58.2	14.6	−58.2
0.8*H*	−33.6	8.4	−14.1	−64.9	16.2	−64.9
H	−116.7	29.2	−107.1	−75.6	18.9	−75.6

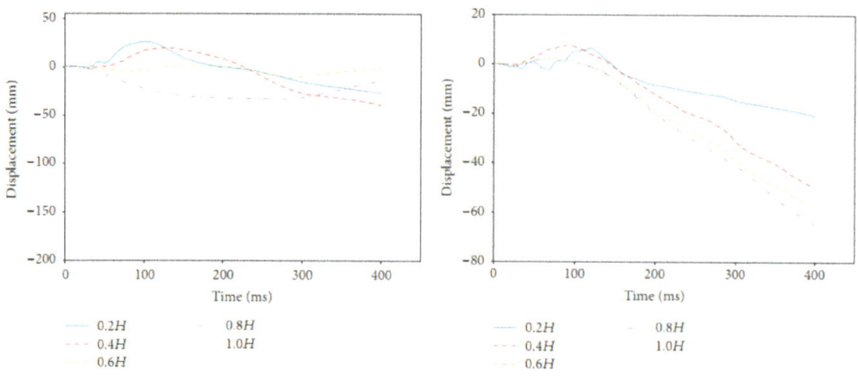

(a) Horizontal displacement of block with respect to the time elapsed (b) Vertical displacement of block with respect to the time elapsed

Figure 8: Block displacement for 600 mm reinforcement spacing.

As shown in Table 8, all the final horizontal displacements were found to be negative, with the largest final horizontal displacement found at the top of the reinforced retaining wall, with a value of −107.1 mm, and with the lowest final horizontal displacement at a height of 0.6H with a value of −1.9 mm. The separation ratio, contrasting the original block size with the maximum displacement of the block, in terms of horizontal displacement, ranged between 6.8 and 29.2%. The final vertical displacements were all found to be negative values, and they increased in magnitude with increasing height of the reinforced retaining wall. The largest final vertical displacement, −75.6 mm, occurred at the top (H) of the reinforced retaining wall, and the separation ratio, contrasting the original block size with the maximum displacement of the block, in terms of vertical displacement, ranged between 5.2 and 18.9%. In Figure 8(a), the movement of the reinforced retaining wall block at 0.2H was found to be dangerous because the reinforced retaining wall at 0.2H was first moved in the direction of the backfill by the initial impact of the vehicle and then it moved again in the direction of the collided vehicle. This movement of the reinforced retaining wall at 0.2 makes the overturning of the structure possible. Furthermore, at this reinforcement spacing, a weaker resistance of the reinforcement was observed because the displacement of the reinforced retaining wall did not recover after the displacement of the reinforced retaining wall increased.

Comparison with Reinforced Retaining Wall Displacements for Different Reinforcement Spacing

Figure 9 shows the absolute value of both the final horizontal and vertical displacements at 0.2H of the reinforced retaining wall for different reinforcement spacings. The final horizontal and vertical displacement increased with an increase in reinforcement spacing.

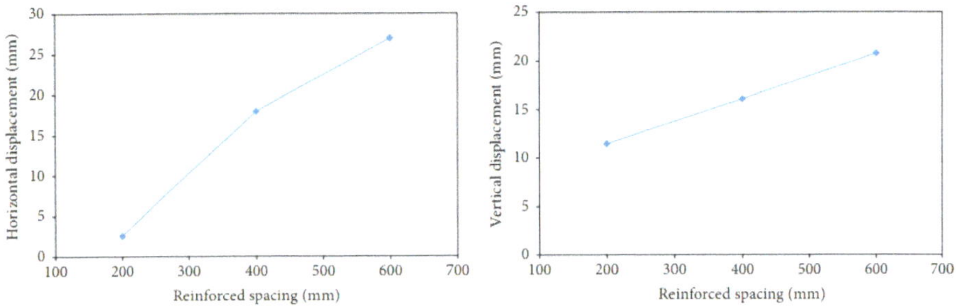

(a) Final horizontal displacement for different reinforcement spacing (b) Final vertical displacement for different reinforcement spacing

Figure 9: Final block displacement as a function of reinforcement spacing.

Effect of Reinforcement at the Top (H) of Reinforced Retaining Wall

As can be deduced from Figure 4, the highest point up to which reinforcement is located is 600 mm from the top of the reinforced retaining wall, for 600 mm reinforcement spacing. Similarly, it is 200 mm from the top of the reinforced retaining wall for 200 mm reinforcement spacing and 400 mm from the top of the reinforced retaining wall for 400 mm reinforcement spacing. As can be seen in Figure 10, the block at the top (H) of the reinforced retaining wall considering the 200 and 400 mm reinforcement spacing more or less returned to the initial position. However, the block at the top (H) of the reinforced retaining wall with 600 mm reinforcement spacing did not recover its original position. This was because the reinforcement was located far from the top of the reinforced retaining wall. Based on this result, it can be deduced that the reinforcement spacing should be within 400 mm to ensure that reinforcement occurs up to the top of the reinforced retaining wall.

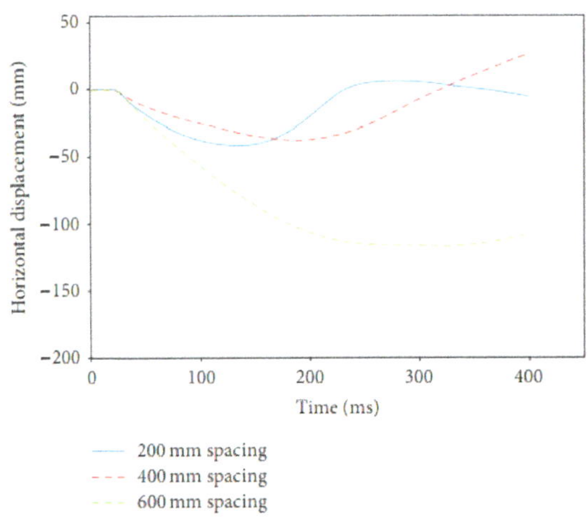

Figure 10: Block displacement at the top (H) of reinforced retaining wall for reinforcement spacing.

Behavior of Backfill for Different Reinforcement Spacing

A reinforced retaining wall serves as a normal retaining wall and is used for bridge abutment (the subbase course of roads) and as a structure surrounding a foundation. The backfill of the wall is a crucial factor in the success of all these functions and therefore requires further consideration because of its effect upon the safety of the structure and because of the irregularity of vehicle running.

Figure 11 shows the variation in the vertical displacement at the top of the backfill with different reinforcement spacing. The analyzed points along the length of the reinforced retaining wall which is L were 0.1L, 0.3L, 0.5L, and 0.8L, where 0.5L is the total length of the reinforcement. For all the reinforcement spacing considered (200, 400, and 600 mm), the 0.1L point was closest to the collision surface of the vehicle, and it was found that the vertical displacement at this point had an invariably positive value, with no large variations for different spacing. The largest vertical displacement was observed at 0.5L, and the vertical displacements were −61.7, −75.5, and −82.0 mm for 200, 400, and 600 mm reinforcement spacing, respectively. Additionally, the settlement also appeared at the points that were not reinforced.

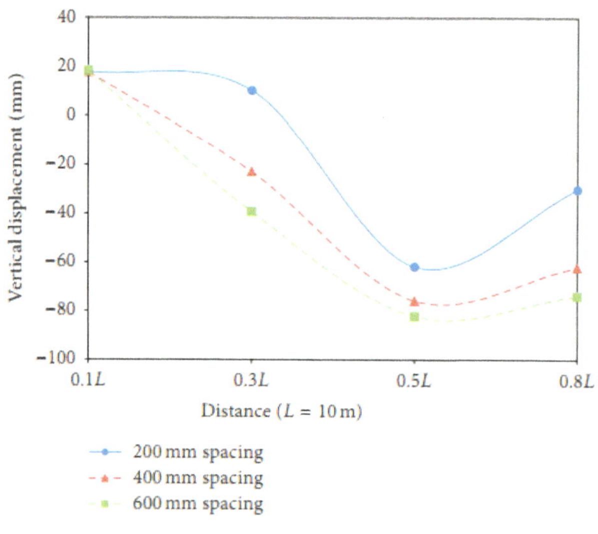

Figure 11: Final vertical displacements along the length of reinforcement for each reinforcement spacing (200, 400, and 600 mm).

CONCLUSION

In this study, the horizontal and vertical displacements of a reinforced retaining wall's front plate, affected by vehicle collision, were examined as functions of reinforcement spacing. The vertical displacement of backfill was also analyzed.

(1) Final horizontal and vertical displacements increased with increase in reinforcement spacing at 0.2H of the reinforced retaining wall height, a point

that is close to the collision contact surface. This means that dense reinforcement spacing allows the reinforced retaining wall to strongly resist external collision load. A decrease in both horizontal and vertical displacement was observed when the interval between reinforcement was decreased. It is possible to conclude that reinforcement spacing is an important resistance factor against external collision load.

(2) When considering 200 mm and 400 mm of reinforcement spacing, the horizontal and vertical displacements were restored in time, after the reinforced retaining wall was impacted by vehicle collision. This means that these reinforcement spacing values (such as 200 mm and 400 mm) can withstand an external collision load. However, there was no restoration for the 600 mm reinforcement spacing. This was deduced to be because reinforcement did not reach up to the top of the reinforced retaining wall. Therefore, it is possible to conclude that the reinforcement spacing has to be considered alongside the total height of the reinforced retaining wall, in order to strengthen the safety of the reinforced retaining wall.

(3) −61.7, −75.5, and −82.0 mm displacements of the top of backfill were examined. It was possible to conclude from these observations that reinforcement affects backfill in its resistance against external collision load and that the nature of the reinforcement is a necessary factor to be considered, when reinforced retaining walls are designed.

REFERENCES

1. V. N. Ghionna, V. Fioravante, and M. Vicari, "Full scale test on a retaining wall non-uniform reinforcements," in Proceedings of the 7th International Conference on Geosynthetics (ICG '02), pp. 279–282, September 2002.

2. D. Leshchinsky and C. Vulova, "Numerical investigation of the effects of geosynthetic spacing on failure mechanisms in MSE block walls," Geosynthetics International, vol. 8, no. 4, pp. 343–365, 2001.

3. H. Liu and M.-S. Won, "Long-term reinforcement load of geosynthetic-reinforced soil retaining walls,"Journal of Geotechnical and Geoenvironmental Engineering, vol. 135, no. 7, pp. 875–889, 2009.

4. M. I. M. Pinto and T. W. Cousens, "Modelling a geotextile-reinforced, brick-faced soil retaining wall,"Geosynthetics International, vol. 6, no. 5, pp. 417–447, 1999.

5. A. W. Stuedlein, T. M. Allen, R. D. Holtz, and B. R. Christopher, "Assessment of reinforcement strains in very tall mechanically stabilized earth walls," Journal of Geotechnical and Geoenvironmental Engineering, vol. 138, no. 3, pp. 345–356, 2012.

6. S. B. A. Mohamed, K.-H. Yang, and W.-Y. Hung, "Finite element analyses of two-tier geosynthetic-reinforced soil walls: comparison involving centrifuge tests and limit equilibrium results," Computers and Geotechnics, vol. 61, pp. 67–84, 2014.

7. K. S. Wong, B. B. Broms, and B. Chandrasekaran, "Failure modes at model tests

of a geotextile reinforced wall," Geotextiles and Geomembranes, vol. 13, no. 6-7, pp. 475–493, 1994.

8. C. Yoo and H.-Y. Jung, "Case history of geosynthetic reinforced segmental retaining wall failure," Journal of Geotechnical and Geoenvironmental Engineering, vol. 132, no. 12, pp. 1538–1548, 2006.
9. M. Borovinšek, M. Vesenjak, M. Ulbin, and Z. Ren, "Simulation of crash tests for high containment levels of road safety barriers," Engineering Failure Analysis, vol. 14, no. 8, pp. 1711–1718, 2007.
10. C. B. Ju, "A study on the embedded depth of guardrail posts," Study of Highway and Transportation Technology Institute in Korea, 2003.
11. Y. Itoh, C. Liu, and R. Kusama, "Dynamic simulation of collisions of heavy high-speed trucks with concrete barriers," Chaos, Solitons and Fractals, vol. 34, no. 4, pp. 1239–1244, 2007.
12. J. Kim, S. Lee, and H. Choi, "Progressive collapse resisting capacity of moment frames with viscous dampers," The Structural Design of Tall and Special Buildings, vol. 22, no. 5, pp. 399–414, 2013.
13. S. K. Tay, B. Lim, and N. G. S. Herng, "Crash impact modelling of security bollard," in Proceedings of the 12th International LS-DYNA Users Conference 13, pp. 1–10, June 2012.
14. W. Wu and R. Thomson, "A study of the interaction between a guardrail post and soil during quasi-static and dynamic loading," International Journal of Impact Engineering, vol. 34, no. 5, pp. 883–898, 2007.
15. K. Ahn, Y. Heo, K. Hong, and M. Ahn, "Behavior of reinforced earth retaining wall with vehicle collision," Journal of the Korean Geoenvironmental Society, vol. 11, no. 6, pp. 39–46, 2010.
16. Ministry of Land, Transport, and Maritime Affairs in Korea, Business Guide for Vehicle Crash Test of Safety Facility of Vehicle Protection, Ministry of Land, Transport and Maritime Affairs in Korea, 2008.
17. Guidelines for Installation and Administration of Load Safety Facility, Ministry of Land, Transport and Maritime Affairs in Korea, 2014.

3
CHAPTER

DYNAMIC ACTIVE EARTH PRESSURES OF THE RETAINING PILES WITH ANCHORS UNDER VEHICLE LOADS

Hong-zhi Qiu, Ji-ming Kong and Ren-chao Wang

INTRODUCTION

In design and construction of the excavation, the calculation problem of earth pressure has been encountered by engineers. There are a lot of deep excavations in center area of the city, where dense urban roads traffic is heavy. In particular, the project of urban road expansion, during construction of the road, keeps working. When foundation pit is close to driving roads, it can cause the damage of retaining structure for underestimating the vehicle vibration on the impact of foundation pit supporting structure (see Figure 1). Usually these typical examples can be commonly found in large- and medium-size cities in China. Therefore, calculating the earth pressure on retaining piles close to road in the design and construction of foundation pit, we must consider the impact of vehicle loads.

Pavement crack

Figure 1: Typical example of pit destruction [16].

Citation: Hong-zhi Qiu, Ji-ming Kong, and Ren-chao Wang, "Dynamic Active Earth Pressures of the Retaining Piles with Anchors under Vehicle Loads," Shock and Vibration, vol. 2016, Article ID 4023827, 8 pages, 2016. doi:10.1155/2016/4023827.

In the past few decades, though the calculation method of earth pressure on retaining structure has been greatly developed, achievements are mainly focused on the dynamic response of the structure subjected to dynamic loads such as earthquake and vehicle loads [1, 2]. Under earthquake loads, the methods have been developed to evaluate the seismic earth pressure on a rigid retaining structure, being known as the pseudostatic method first used by Okabe [3] and Mononobe and Matsuo [4]. Pseudostatic method is an extension of the Coulomb sliding wedge theory, which was later recognized as famous Mononobe-Okabe method [5]. Steedman and Zeng [6] proposed a pseudodynamic method, which assumes that the walls and high seismic acceleration versus time were sinusoidal variation, and the shear modulus and shear wave velocity are also identified as a limited value; the study shows that the pseudodynamic method is closed actually. Then, Zeng and Steedman [7] also show that the earth pressure in pseudodynamic method is in agreement with the centrifugal modeling results, confirming that the proposed method is correct.

Pseudodynamic method is used to calculate the earth pressure on retaining walls under earthquake; the results are more accurate compared to the static method [8–12]. Study on supporting structure under the vehicle load is mainly in the dynamic characteristics of the supporting structures [13, 14] and the stability of supporting structures [15]. However, in the calculation method of earth pressure behind the supporting piles, effects of vehicle vibration characteristics are rarely taken into account. In addition, there is no consideration of the constraint of anchor structures in solution of earth pressure on retaining piles.

In the past, solving the earth pressure on supporting structure is based on static method [17, 18], and vehicle loads are simplified equal static load; results of calculation show that the error is relatively large and cannot reflect the soil pressure in practice. Thus, in this work, a distribution of earth pressure on supporting structure under vehicle load is presented; we have considered the contribution of anchor structures on earth pressure on piles. Moreover, effects of a wide range of parameters like vibration acceleration coefficient, wall friction angle, soil friction angle, and the weight of the soil on active earth pressure have been studied. Finally, comparison of the earth pressure is calculated by M-O method and present study and test data; a suggestion has been given to calculate earth pressure on pile under vehicle loads.

METHOD OF ANALYSIS

Response Acceleration

The growth of heavy vehicle loads poses a threat to the safety of infrastructure. Dynamic stress and acceleration response analysis is important for design and assessment of structure [19, 20]. Acceleration induced by vehicle loads was obtained by experiment [21, 22]. The experimental results show that characters measured acceleration response versus time is similar to acceleration induced by earthquake. Therefore, this paper assumes that the vehicle generates shear and primary waves.

Pseudodynamic method assumes that the rock mass shear modulus G does not change over vertical position of the slope; namely, G is a constant. Supporting piles at depth z by horizontal and vertical vibration accelerations is defined as $a_h(z, t)$ and $a_V(z, t)$, respectively, where z is depth below the surface of the soil vibration acceleration at time t for the following equation [6]:

$$a_h(z,t) = k_h g \cdot \sin\left[\omega\left(t - \frac{z}{V_s}\right)\right],$$

$$a_V(z,t) = k_V g \cdot \sin\left[\omega\left(t - \frac{z}{V_p}\right)\right], \tag{1}$$

where k_h is the horizontal vibration acceleration coefficient, k_V is the vertical vibration acceleration coefficients, z is the depth of the underground, ω is the angular frequency of the wave vibration, and V_s and V_p are the velocity of shear wave and vertical wave, respectively:

$$V_s = \sqrt{\frac{G}{\rho}},$$

$$V_p = \sqrt{\frac{G(2-2\mu)}{\rho(1-2\mu)}}, \tag{2}$$

where G is the shear modulus of soil and ρ and μ are the density of pile soil and Poisson's ratio, respectively.

For the majority of geological material, $V_p/V_s = 1.87$ [23], and vibration cycle is $T = 2\pi/\omega = 4H/V_s$ [24].

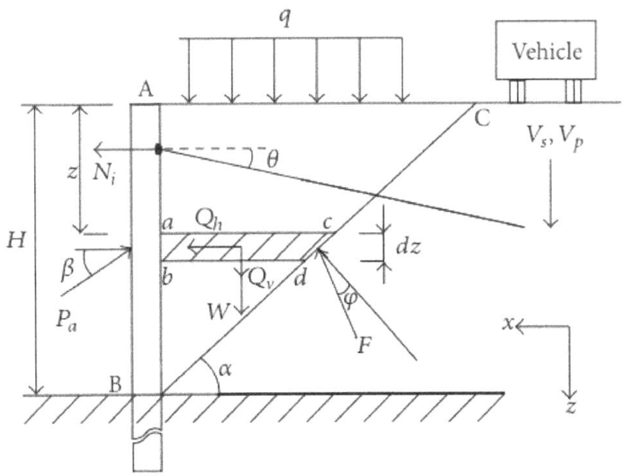

Figure 2: Model for computation earth pressure on piles under vehicle loads.

Analytical Model.

Based on the character of the research problem, the active earth pressure behind retaining piles has been analyzed by using pseudodynamic approach under vehicle loads. According to practical engineering problems, analysis model is shown in Figure 2. The soil behind the piles from an equilibrium state to another equilibrium state in Figure 3 and the position *AB* moves to a new location *A*' *B*' , so a slight angle δ is generated; at the same time, the pile shows a slight lateral deformation Δl. We assume that the lateral deformation of pile is approximately equal to the elongation of anchor structure, and the tensile force of anchor is ΔN determined by the elongation of anchor, which distributed continuously with the depth:

$$\Delta l = (H - z)\,\theta, \tag{3}$$

where H is a length of cantilever retaining piles.

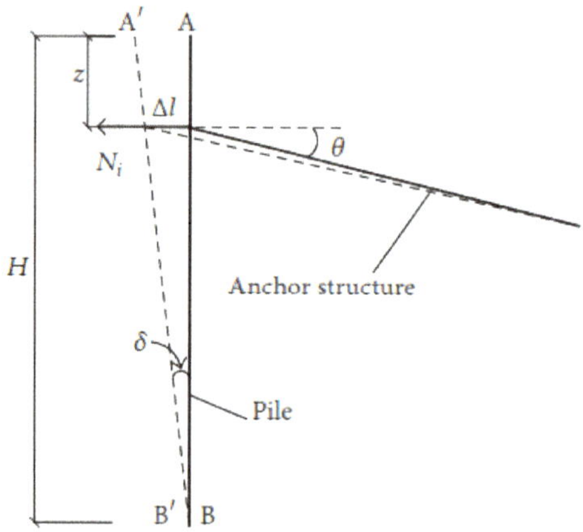

Figure 3: Lateral deformation of pile-anchor support structure.

According to the principle of force equilibrium, the tensile force acting on the piles in depth z is given by

$$\Delta N = \Delta l E A_a = (H - z)\,\theta E A_a, \tag{4}$$

where E is elastic modulus of anchor structure and A_a is sectional area of the anchor structure. Tension generated by anchor structure is assumed to be evenly distributed along the length of pile, and then the total force can be expressed as

$$N = \int_0^H \Delta N\, dz = \frac{1}{2} H^2 \theta E A_a. \tag{5}$$

From Figure 2 in the limit equilibrium soil wedge ABC, taking dz thickness analysis unit body as an object,

$$\overline{ac} = (H - z)\cot\alpha,$$

$$\overline{bd} = (H - z - dz)\cot\alpha. \tag{6}$$

Then, the area of the horizontal bar dz is

$$S_{acdb} = \frac{1}{2}\left(\overline{bd} + \overline{ac}\right)dz. \tag{7}$$

The quality of the horizontal bar dz is

$$\Delta m = \rho S_{acdb} = \rho(H - z)\cot\alpha\, dz. \tag{8}$$

The weight of the soil wedge is

$$W = \int_0^H \rho g(H - z)\cot\alpha\, dz = \frac{1}{2}\gamma H^2\cot\alpha, \tag{9}$$

where γ is the unit weight of the soil, $\gamma = \rho g$.

The total horizontal inertia force acting within the failure zone is given by

$$\begin{aligned} Q_h(z,t) &= \int_0^H \Delta m a_h(z,t) \\ &= \frac{\lambda_s \gamma k_h \cot\alpha}{2\pi} \\ &\quad \cdot \left[\frac{\lambda_s}{2\pi}(\sin\omega t - \sin\omega\eta) - H\cos\omega t\right], \end{aligned} \tag{10}$$

where λ_s is the transverse wavelength $\lambda_s = V_s/f$ and the angular frequency $\omega = 2\pi f$, making $\eta = t - H/V_s$. Similarly, the total vertical inertial force acting on the failure wedge is given by

$$\begin{aligned} Q_V(z,t) &= \int_0^H \Delta m a_V(z,t) \\ &= \frac{\lambda_p \gamma k_V \cot\alpha}{2\pi}\left[\frac{\lambda_p}{2\pi}(\sin\omega t - \sin\omega\xi) - H\cos\omega t\right], \end{aligned} \tag{11}$$

where λ_p is the longitudinal wave $\lambda_p = V_p/f$, $V_p/\omega = \lambda_p/2\pi$, making $\xi = t - H/V_p$. According to the equilibrium of the forces on piles in Figure 2, the total active thrust can be obtained, so Pa can be expressed as follows:

$$P_a = \frac{\cos(\alpha-\varphi)Q_h + \sin(\alpha-\varphi)Q_V + \sin(\alpha-\varphi)W}{\cos(\alpha-\varphi-\beta)} + \frac{\cot\alpha\sin(\alpha-\varphi)qH + \cos(\alpha-\varphi+\theta)N}{\cos(\alpha-\varphi-\beta)}. \tag{12}$$

Assuming that the coefficient of active earth pressure is K_a,

$$K_a = \frac{2P_a}{\gamma H^2}. \tag{13}$$

Substituting for Q_h and Q_V in (13), an expression for K_a in terms of Q_h, Q_V, W, and N can be derived:

$$K_a = \frac{\cos(\alpha-\varphi)\cot\alpha k_h}{2\pi^2\cos(\alpha-\varphi-\beta)}\frac{TV_s}{H}a_1 + \frac{\sin(\alpha-\varphi)\cot\alpha k_V}{2\pi^2\cos(\alpha-\varphi-\beta)}\frac{TV_p}{H}a_2 + \frac{\cot\alpha\sin(\alpha-\varphi)}{\cos(\alpha-\varphi-\beta)} + \frac{2q\cot\alpha\sin(\alpha-\varphi)}{\gamma H\cos(\alpha-\varphi-\beta)} + \frac{\cos(\alpha-\varphi+\theta)\theta EA_a}{\gamma\cos(\alpha-\varphi-\beta)},$$

$$a_1 = \frac{TV_s}{H}\left[\sin 2\pi\left(\frac{t}{T}\right) - \sin 2\pi\left(\frac{t}{T} - \frac{H}{TV_s}\right)\right] - 2\pi\cos 2\pi\left(\frac{t}{T}\right),$$

$$a_2 = \frac{TV_p}{H}\left[\sin 2\pi\left(\frac{t}{T}\right) - \sin 2\pi\left(\frac{t}{T} - \frac{H}{TV_p}\right)\right] - 2\pi\cos 2\pi\left(\frac{t}{T}\right). \tag{14}$$

From (14) it can be analyzed that active earth pressure coefficient is the function of α, t/T, H/TV_s, and H/TV_V. In most geological materials H/TV_s and H/TV_p values are 0.25 and 0.134, respectively. And then we can take the partial derivative with respect to (z, t) as distribution of earth pressure with depth:

$$
\begin{aligned}
p_a(z,t) &= \frac{\partial P_a(z,t)}{\partial z} \\
&= \frac{\cos(\alpha-\varphi)\,TV_s}{\cos(\alpha-\varphi-\beta)\,2\pi}\gamma k_h(\cot\alpha)\,b_1 \\
&\quad + \frac{\sin(\alpha-\varphi)\,TV_p}{\cos(\alpha-\varphi-\beta)\,2\pi}\gamma k_V(\cot\alpha)\,b_2 \\
&\quad + \frac{\sin(\alpha-\varphi)\cot\alpha q}{\cos(\alpha-\varphi-\beta)} \\
&\quad + \frac{\sin(\alpha-\varphi)\,z\gamma\cot\alpha}{\cos(\alpha-\varphi-\beta)} \\
&\quad + \frac{\cos(\alpha-\varphi+\theta)\,z\theta EA_a}{\cos(\alpha-\varphi-\beta)},
\end{aligned}
$$

$$
b_1 = \left[\cos 2\pi\left(\frac{t}{T}-\frac{z}{TV_s}\right)-\cos 2\pi\left(\frac{t}{T}\right)\right],
$$

$$
b_2 = \left[\cos 2\pi\left(\frac{t}{T}-\frac{z}{TV_p}\right)-\cos 2\pi\left(\frac{t}{T}\right)\right]. \tag{15}
$$

According to (15), dynamic active earth pressures on piles for foundation pit contain three parts:

$$
p_a(z,t) = p_{ad} + p_{as} + p_{at},
$$

$$
\begin{aligned}
p_{ad} &= \frac{\cos(\alpha-\varphi)\,TV_s}{\cos(\alpha-\varphi-\beta)\,2\pi}\gamma k_h(\cot\alpha)\,b_1 \\
&\quad + \frac{\sin(\alpha-\varphi)\,TV_p}{\cos(\alpha-\varphi-\beta)\,2\pi}\gamma k_V(\cot\alpha)\,b_2,
\end{aligned}
$$

$$
\begin{aligned}
p_{as} &= \frac{\cot\alpha\sin(\alpha-\varphi)\,q}{\cos(\alpha-\varphi-\beta)} \\
&\quad + \frac{\sin(\alpha-\varphi)}{\cos(\alpha-\varphi-\beta)}z\gamma\cot\alpha,
\end{aligned}
$$

$$
p_{at} = \frac{\cos(\alpha-\varphi+\theta)\,z\theta EA_a}{\cos(\alpha-\varphi-\beta)}. \tag{16}
$$

p_{ad} is the dynamic earth pressure created by vehicle loads; p_{as} is the static earth pressure created by the weight of soils and the static load on the ground; p_{at} is the tensile force produced by anchor structure.

Computation of Rupture Angle (α).

It can be shown from (15) that earth pressure distribution function (z, t) is related to the following parameters: φ, β, γ, k_h, k_V, and f_a. And test shows that the soil pressure after pile varies with time and different depth z. When the sliding mass attains limiting equilibrium condition, the active earth pressure after pile (z, t) reached a maximum. Thus, the angle of sliding surface inclination and also the angle of rupture can be obtained by partial differential equation $\partial P_a(z, t)/\partial\alpha = 0$. After simplifying,

$$\begin{aligned}
&\left[\frac{-\sin\beta\cot\alpha}{\cos^2(\alpha-\varphi-\beta)} - \frac{\cos(\alpha-\varphi)}{\cos(\alpha-\varphi-\beta)\sin^2\alpha}\right]\frac{\lambda_s\gamma k_h}{2\pi} \\
&\quad\cdot m_1 + \left[\frac{\cos\beta\cot\alpha}{\cos^2(\alpha-\varphi-\beta)}\right. \\
&\quad\left. - \frac{\sin(\alpha-\varphi)}{\cos(\alpha-\varphi-\beta)\sin^2\alpha}\right]\frac{\lambda_V\gamma k_V}{2\pi}m_2 \\
&\quad + \frac{\gamma H^2}{2}\left[\frac{\cos\beta\cot\alpha}{\cos^2(\alpha-\varphi-\beta)}\right. \\
&\quad\left. - \frac{\sin(\alpha-\varphi)}{\cos(\alpha-\varphi-\beta)\sin^2\alpha}\right] + qH\left[\frac{\cot\alpha\cos(\alpha-\varphi)}{\cos(\alpha-\varphi-\beta)}\right. \\
&\quad\left. - \frac{\sin(\alpha-\varphi)}{\cos(\alpha-\varphi-\beta)\sin^2\alpha}\right] - \frac{\sin(\beta+\theta)H^2\delta EA_a}{2\cos^2(\alpha-\varphi-\beta)} \\
&\quad = 0, \\
m_1 &= \frac{\lambda_s}{2\pi}\left[\sin 2\pi\left(\frac{t}{T}\right) - \sin 2\pi\left(\frac{t}{T} - \frac{H}{TV_s}\right)\right] - H \\
&\quad\cdot\cos 2\pi\left(\frac{t}{T}\right), \\
m_2 &= \frac{\lambda_V}{2\pi}\left[\sin 2\pi\frac{t}{T} - \sin 2\pi\left(\frac{t}{T} - \frac{H}{TV_p}\right)\right] - H \\
&\quad\cdot\cos 2\pi\left(\frac{t}{T}\right).
\end{aligned} \tag{17}$$

Equation (17) is a function of the time t and the inclination angle α of the slide surface, abbreviated as $(t, \alpha) = 0$. Therefore, the angle of rupture will be obtained by drawing the soil pressure distribution curve since corresponding angle α_i at any time will all work out.

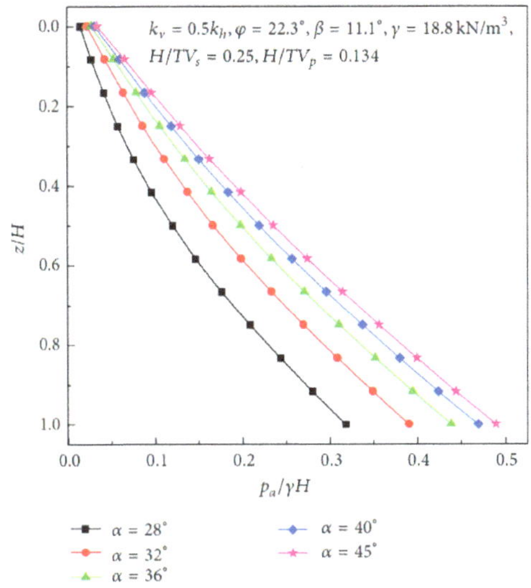

Figure 4: Active earth pressure distribution for different values of rupture angle, *α*.

RESULTS AND DISCUSSION

Effect of Rupture Angle (*α*)

Figure 4 shows the distributions of active earth pressure for values of rupture angle α with $k_V = 0.5k_h$, $k_h = 0.05$, $\varphi = 22.3°$, $\beta = \varphi/2$, $\gamma = 18.8$ kN/m³, $H/TV_s = 0.25$, and $H/TV_p = 0.134$. With the increasing of rupture angle α, active earth pressure increased. Plots show that when α change from 28° to 32°, the active earth pressure increases by 16.29 kN; namely, when each degree increases in rupture angle α, the active earth pressure increased by 4.07 kPa.

Similarly, when α change from 32° to 36°, a total of the active earth pressure increases by 10.81 kPa and the average of active earth pressure increased by 2.7 kPa for each degree. The value of increment in active earth pressure decreased with increase at the same angle (α). Therefore, the greater the rupture angle, the smaller the impact range on the earth pressure. The dotted line represents the dynamic earth pressure produced by vehicle loads; the solid line represents the static earth pressure in Figure 5. Results show that the value of rupture angle decreases, with increase in proportion of dynamic earth pressure. Moreover, the greater the rupture angle, the smaller the dynamic earth pressure.

In Figure 6, the solid line represents the dynamic earth pressure produced by the horizontal inertia force, and the dotted line represents the dynamic earth pressure produced by the vertical inertial force; comparing the two values, results show that the vertical inertial force has a marginal effect on the dynamic earth pressure and thus contributions to dynamic earth pressure are mainly horizontal inertia force.

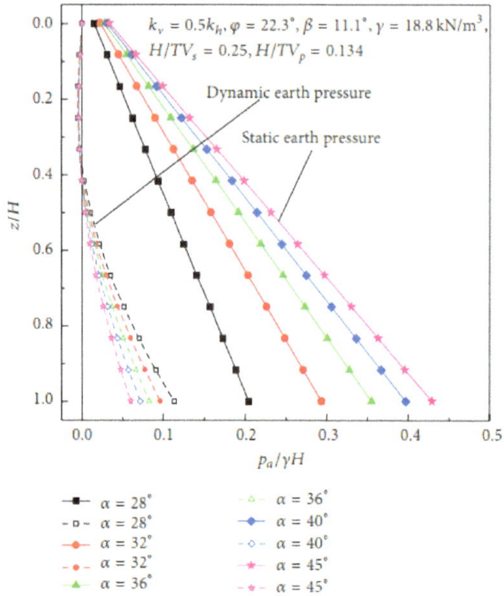

Figure 5: Comparisons of static and dynamic soil pressure.

Effect of k_h and k_V.

According to the definition of the acceleration coefficient, k_h can indirectly reflect the value of inertial force by dynamic loads.

Figure 7 shows a plot of the distributions of active earth pressure for values of k_h with $k_V = 0.5k_h$, $\varphi = 22.3°$, $\beta = \varphi/2$, $\gamma = 18.8$ kN/m^3 , $H/TV_s = 0.25$, and $H/TV_p = 0.134$.The plot shows that, at more than half the height of the pile, the impact on the earth pressure is very small and active earth pressure decreases with the increase of coefficient, k_h. At less than half the height of the pile, earth pressure is greatly influenced by k_h and active earth pressure increases with the acceleration coefficient increase.

Effect of Soil Friction Angle (φ).

Figure 8 shows the distributions of active earth pressure for values of soil friction angle φ with $k_V = 0.5k_h$, $k_h = 0.05$, $\beta = \varphi/2$, $\gamma = 18.8$ kN/m^3 , $H/TV_s = 0.25$, and $H/TV_p = 0.134$. When φ change from 18° to 26° , active earth pressure changes from 152.67 kPa to 105.69 kPa, which decreases by about 30.77%. In other words, each degree increased in the soil friction angle φ, and then active earth pressure decreases by 5.9 kPa.

Effect of Wall Friction Angle (β).

Figure 9 shows the distributions of active earth pressure for values of wall friction angle β with $k_V = 0.5k_h$, $k_h = 0.05$, $\varphi = 20°$, $\gamma = 18.8$ kN/m^3 , $H/TV_s = 0.25$, and $H/TV_p = 0.134$. Active earth pressure shows relatively marginal decrease with the increase in wall friction angle β. When β change from 0° to $\varphi/2$, the active earth pressure decreases by 3.48%, and when β change from 0° to φ, the active earth pressure decreases by 3.65%.

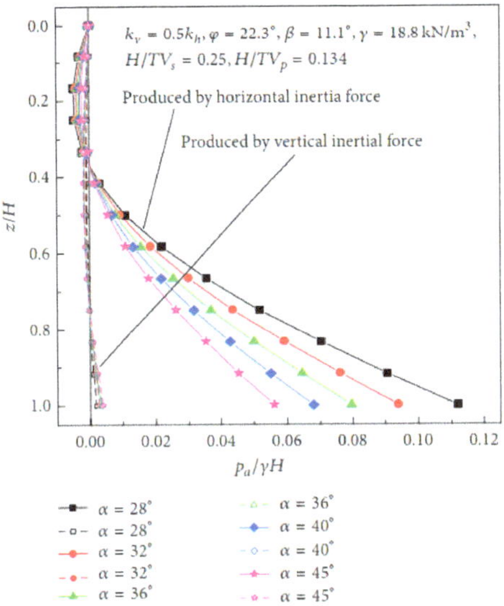

Figure 6: Comparisons of dynamic earth pressures produced by the horizontal inertia force and the vertical inertial force.

From (12), it can be learned that the generation of active earth pressure on retaining structure consists of three parts, including inertia force generated by vehicle loads, tension of anchor, and the gravity of soil wedge. The wall friction angle has no effect on the total earth pressure. In addition, the wall friction angle in the formula is only a denominator, calculating the distribution of earth pressure with depth. The denominator is a function of three angles, one of the angles changes, and the others change as well. Therefore, the relation between all angles is to ensure that the denominator is positive and the effect of the wall friction angle β is not worth mentioning.

Comparison of Results.

In order to certify that present study method is effective, Figure 10 shows the comparison of the test data and the active earth pressure calculated by Mononobe-Okabe method and present study with $k_V = 0.5k_h$, $k_h = 0.04$, $\varphi = 22.3°$, $\beta = \varphi/2$, $\gamma = 18.8$ kN/m^3, $H/TV_s = 0.25$, and $H/TV_p = 0.134$; the above parameters are obtained by experience.

It is evident from Figure 10 that active earth pressure calculated by present study is close to the test data; result obtained by Mononobe-Okabe method is relatively large. What is more, distributions of active earth pressure calculated by present study and test data show clearly nonlinear vibration of vehicle loads. It is also clear from Steedman and Zeng [6, 7] that the dynamic earth pressure distribution along the height is nonlinear.

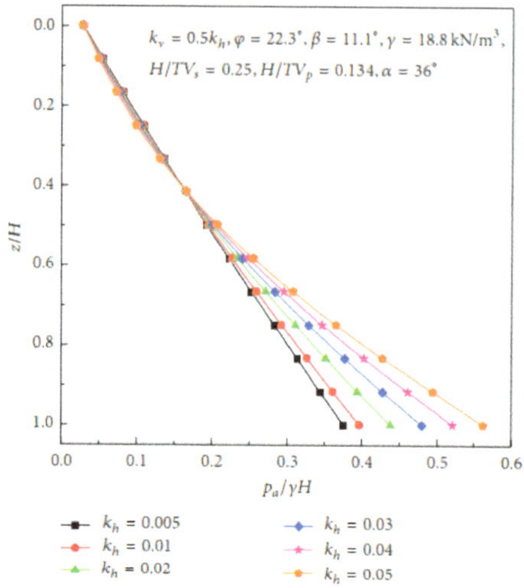

Figure 7: Active earth pressure distribution for different values of k_h with $k_v = 0.5k_h$.

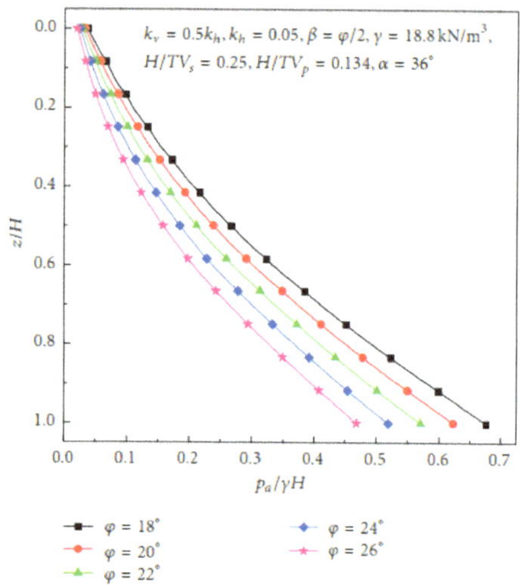

Figure 8: Active earth pressure distribution for different values of soil friction angle, φ.

CONCLUSIONS

Pile-anchor retaining structure is the type of deep foundation pit retaining system most frequently used in China. In this work, based on the pseudodynamic method, a solution was presented for calculating active earth pressures on piles in deep pits. The proposed solution incorporates vibration of vehicle loads, retaining structure deformations and material properties. The following are the main conclusions of the study:

(1) In present method, by considering the effect of vehicle loads and the constraint of anchor structures, the solution to calculate the active earth pressure on piles under vehicle loads is obtained. The paper presents effects of parameters on earth pressure, such as rupture angle, horizontal and vertical vibration coefficient, soil friction angle, and wall friction angle.

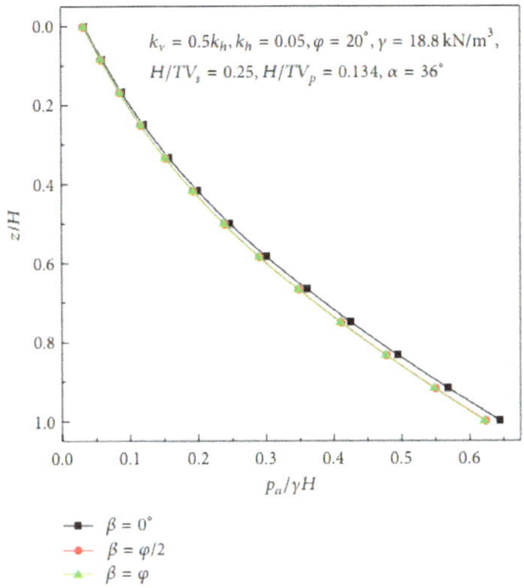

Figure 9: Active earth pressure distribution for different values of wall friction angle, β.

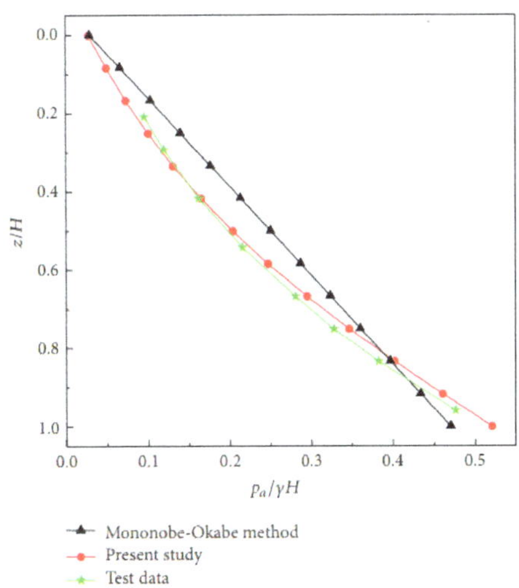

Figure 10: Typical comparison results for $k_h = 0.04$, $k_V = 0.02$, $\varphi = 22.3^\circ$, $\beta = 11.1^\circ$, and $\gamma = 18.8$ kN/m^3.

(2) The results of present analysis show that the earth pressure increases with the increase in rupture angle, α, and horizontal and vertical vibration coefficient, k_h and k_V, which have a great influence on the earth pressure on piles. The active earth pressure is highly sensitive to the friction angle of the soil, φ, and comparatively less sensitive to the wall friction angle, β.

(3) The typical results of distribution of the earth pressure on piles show the high nonlinearity. Under the influence of many factors, the distribution of dynamic earth pressure generated by vehicle loads with depth is nonlinear, which have a great contribution to the active earth pressure on piles. In comparison of the test data and the results calculated by present study and Mononobe-Okabe method, the results show that the earth pressure gained by the proposed method coincides well with the test data.This work provides a method to calculate earth pressure on piles required for the design purpose.

ACKNOWLEDGMENTS

This research was supported by National High Technology Research and Development Program of China ("863" Project, no. 2012AA121302). In addition, NetEase News provided image information for this work.

REFERENCES

1. K. S. Subba Rao and D. Choudhury, "Seismic passive earth pressures in soils," Journal of Geotechnical and Geoenvironmental Engineering, vol. 131, no. 1, pp. 131–135, 2005.
2. A.-H. Soubra, "Static and seismic passive earth pressure coefficients on rigid retaining structures,"Canadian Geotechnical Journal, vol. 37, no. 2, pp. 463–478, 2000.
3. S. Okabe, "General theory of earth pressure," Journal of Japan Society of Civil Engineers, vol. 12, no. 1, 1926 (Japanese).
4. N. Mononobe and H. Matsuo, "On the determination of earth pressure during earthquakes," inProceedings of the World Engineering Conference, vol. 9, pp. 177–185, Tokyo, Japan, 1929.
5. R. V. Whitman, "Seismic design of earth retaining structures," in Proceedings of the 2nd International Conference on Recent Advances in Geotechnical Earthquake Engineering and Soil Dynamics, pp. 1767–1778, St. Louis, Mo, USA, March 1991.
6. R. S. Steedman and X. Zeng, "The influence of phase on the calculation of pseudo-static earth pressure on a retaining wall," Geotechnique, vol. 40, no. 1, pp. 103–112, 1990.
7. X. Zeng and R. S. Steedman, "On the behaviour of quay walls in earthquakes," Geotechnique, vol. 43, no. 3, pp. 417–431, 1993.
8. D. Choudhury and S. S. Nimbalkar, "Pseudo-dynamic approach of seismic active

earth pressure behind retaining wall," Geotechnical and Geological Engineering, vol. 24, no. 5, pp. 1103–1113, 2006.

9. Z.-K. Wang, T.-D. Xia, and W.-Y. Chen, "Pseudo-dynamic analysis for seismic active earth pressure behind rigid retaining wall," Journal of Zhejiang University (Engineering Science), vol. 46, no. 1, pp. 46–51, 2012 (Chinese).
10. P. Ghosh, "Seismic passive earth pressure behind non-vertical retaining wall using pseudo-dynamic analysis," Geotechnical and Geological Engineering, vol. 25, no. 6, pp. 693–703, 2007.
11. S. Ghosh, "Pseudo-dynamic active force and pressure behind battered retaining wall supporting inclined backfill," Soil Dynamics and Earthquake Engineering, vol. 30, no. 11, pp. 1226–1232, 2010.
12. A. Azad, S. S. Yasrobi, and A. Pak, "Seismic active pressure distribution history behind rigid retaining walls," Soil Dynamics and Earthquake Engineering, vol. 28, no. 5, pp. 365–375, 2008.
13. C. Lin, Y.-F. Luo, and S.-P. Wang, "Monitoring and effect of moving load to foundation excavation support configuration," Journal of Wuhan University of Technology, vol. 29, no. 11, pp. 112–114, 2007 (Chinese).
14. J. C. Yue, H. Z. Qiu, and L. J. Zhang, "Analysis on dynamic response of the foundation pit supporting structure under traffic loads," Chinese Journal of Underground Space and Engineering, vol. 9, no. 6, pp. 1320–1325, 2013 (Chinese).
15. X. D. Zhang, C. G. Zhang, and J. S. Liu, "Analysis on stability of deep foundation pit supporting structure under traffic load," The Chinese Journal of Geological Hazard and Control, vol. 22, no. 2, pp. 125–129, 2011 (Chinese).
16. Netease news, "The supporting structure of pit fracture and the subway construction site collapsed in Wuhan city, China," 2012, http://news.163.com/photoview/00AN0001/30678.html.
17. J.-J. Wang and H.-J. Chai, "Active earth pressure induced by saturated sub grade under vehicle load,"Chinese Journal of Geotechnical Engineering, vol. 30, no. 3, pp. 372–378, 2008 (Chinese).
18. Specifications for Design of Highway Asphalt Pavement (JTGD50-2006), China Communication Press, Beijing, China, 2006 (Chinese).
19. H. Zhang, X. Xie, J. Q. Jiang, and M. Yamashita, "Assessment on transient sound radiation of a vibrating steel bridge due to traffic loading," Journal of Sound and Vibration, vol. 336, pp. 132–149, 2015.
20. Y. L. Xu, Q. Li, D. J. Wu, and Z. W. Chen, "Stress and acceleration analysis of coupled vehicle and long-span bridge systems using the mode superposition method," Engineering Structures, vol. 32, no. 5, pp. 1356–1368, 2010.
21. H. Aied, A. González, and D. Cantero, "Identification of sudden stiffness changes in the acceleration response of a bridge to moving loads using ensemble empirical mode decomposition," Mechanical Systems and Signal Processing, vol. 66-67, pp. 314–338, 2016.

22. H. Z. Qiu, J. M. Kong, R. C. Wang, et al., "Analyzing the effects of active earth pressure on supporting piles under vehicle loads," Chinese Journal of Geotechnical Engineering, 2015 (Chinese).

23. D. Choudhury and S. Nimbalkar, "Seismic passive resistance by pseudo-dynamic method," Geotechnique, vol. 55, no. 9, pp. 699–702, 2005.

24. S. L. Kramer, Geotechnical Earthquake Engineering, Prentice Hall, Upper Saddle River, NJ, USA, 1990.

4
CHAPTER

LATERAL EARTH PRESSURE BEHIND WALLS ROTATING ABOUT BASE CONSIDERING ARCHING EFFECTS

Dong Li, Wei Wang, and Qichang Zhang

INTRODUCTION

Prediction of earth pressure is an important subject of research for design of retaining structures in prone rupture region. Rupture mechanism in the backfill behind a retaining structure should take into account the mode of wall movement. It is common in practice that movements of retaining wall are considered in terms of translation (T), rotation about base (RB), and rotation about top (RT). Conventional earth pressure theories are brought forward only for retaining structures with rigid free translation, such as methods modeled by Coulomb [1], Rankine [2], Khajehzadeh et al. [3], Xu [4], and differential slice technique Wang [5]. In contrast to the usual methods, the research needs more light on prediction of lateral earth pressure on a rotating wall. Terzaghi [6] conducted several large-scale model tests to recognize the importance of walls rotation. Despite the fail in measuring the actual stresses against the wall, he arrived at his conclusions by measuring the total thrust by load cells. Some researches applied pseudodynamic method to analyze the pressure on the retaining wall, such as [7–10]. However, all of them failed to take into account the arching effects in the soil. James and Bransby [11] investigated the passive failure of an initially vertical plane which was rotated about its base into a mass of dry sand with an unloaded horizontal surface. Although the results showed uniformity between the predicted principal compressive stress directions and the observed principal compressive strain increment direction, this theory only focuses on the failure mechanism of slipping plane and ignores other boundary conditions. Small-scale model of retaining walls (e.g., [12, 13]) and field observations [14] have enhanced the understanding of lateral earth pressure problems of rotation. Chang used a modified Coulomb's solution of active pressure to analyze the earth pressure distribution

Citation: Dong Li, Wei Wang, and Qichang Zhang, "Lateral Earth Pressure behind Walls Rotating about Base considering Arching Effects," Mathematical Problems in Engineering, vol. 2014, Article ID 715891, 7 pages, 2014. doi:10.1155/2014/715891.

due to different wall movements [15]. Recently, Song and Zhang [16] revealed the formation mechanism of earth pressures against rigid retaining wall with RT and RB mode. Nevertheless, soil arching effects should be taken into account to predict earth pressure distributions for application of advanced developments.

Granular matters exhibit some unique character, such as arching effects, which inspire the interest of a lot of scholars [17–20]. Soil arching behavior develops when the earth pressure transfers from the yielding part of soil to the adjacent part. Terzaghi [21] used trap door tests to explain how pressure transferred from yielding parts of a soil mass to adjoined stationary; nonyielding parts led to the formation of an arching zone. Following this pioneering study, numerous works have been carried out to investigate the arching effect in various engineering problems. Handy [22] deduced the distribution of lateral earth pressure behind a retaining wall by assuming that the curve describing the minor principal stress is a catenary. Later, Harrop-Williams [23] theoretically derived that the shape of the arch can be approximated by a circular arc to cohere with the assumption of constant stresses along the arch. Paik and Salgado [24] proposed a new formulation for calculating the active earth pressure on a rigid retaining wall undergoing horizontal translation, which took into account the arching effects that occurred in the retained soil mass. Jiang et al. [25] theoretically analyzed the shape of the minor principal stress arch. Using finite element method, Potts and Zdravkovic [26] reviewed the nature of the soil arching that developed in the retained soil mass. Nadukuru and Michalowski [27] applied discrete element method for calculation of active loads on retaining structures with different movements. The results showed that arching and modes of wall movements appear to be the primary cause affecting the pressure distribution.

Motivated by the above literature review, this paper focuses on an analytical method for the earth pressure distribution on a rigid wall rotating about base (RB) with consideration of the arching effects. First, it infers suitable lateral earth coefficients K assuming that the arch forms a part of circular arc in the soil. Then using differential slice technique as the framework of analysis, a reasonable formula is proposed for RB. It adopts a simple concept of average friction angle of soil to relate the mobilized shearing resistance to wall displacement. At last, comparisons are made between the prediction of proposed method and existing experiments results.

BASIC THEORY

As Terzaghi alluded, arching may play a role in distribution of pressures on retaining structures. Whereas the term arching has been accepted in the geotechnical literature, the concept does not relate to the formation of a physical arch but rather a distribution of pressures for which stiffer components of the system attract more loads. According to Terzaghi's theory, a column of soil is assumed to move downwards under the influence of its own weight and is restrained by friction along the boundary of this falling column and the surrounding soil mass. In a similar way to Handy's method, it is to substitute a circular arc describing the path of the minor principal stress to mimic soil arching effects [24].

LATERAL EARTH PRESSURE OF COEFFICIENTS

The following process is assumed: when the rotation is small, a similar deformation pattern characterized by quasi-rupture surfaces will uniformly distribute in the zone that has not yet run up to the limiting condition but with the shearing resistance only partly activated along these planes, as well as along the soil-wall interface in the same zone. With reference to walls rotating about base, an earlier discussion has indicated that rupture surfaces appear gradually from the upper part of the wall to the base. Figure 1 shows the condition behind a retaining wall after vertical settling of the fill occurs due to the wall movement. For portion of the wall, the shear resistance in soil is just partially mobilized and the friction angle of the soil should be the mean value, $\bar{\phi}$. For the slipping plane in angle of α with the horizontal, it is $\pi/4+\bar{\phi}/2$ for active case owing to Rankine theory. With respect to horizontal backfill retained by a perfectly smooth wall, the earth pressure on the wall is the principal stress and the lateral pressure coefficient assumes a well-known expression [24]. Consider

$$K_a = \frac{1}{N} = \frac{\sigma_3}{\sigma_1} = \tan^2\left(\frac{\pi}{4} - \frac{\bar{\phi}}{2}\right), \tag{1}$$

where N, which can be deduced from Mohr circle, as shown in Figure 2, is

$$N = \frac{\sigma_1}{\sigma_3} = \frac{1+\sin\bar{\phi}}{1-\sin\bar{\phi}} = \tan^2\left(\frac{\pi}{4} + \frac{\bar{\phi}}{2}\right). \tag{2}$$

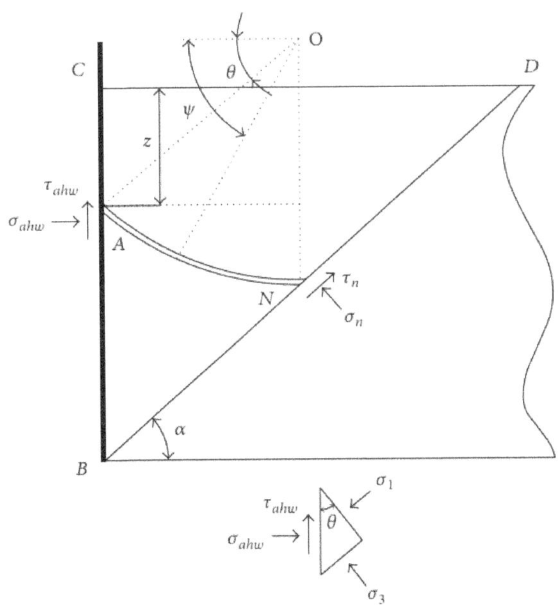

Figure 1: Trajectory of minor principal stress.

However, in field, the direction of the principal stress on the wall is at a slight tilt in view of the wall friction. The arch shown in Figure 1 describes the direction of the minor principal stress in virtue of Handy's theory and also describes the soil enclosed in a region where there are no shearing stresses on the trajectory of the region. It is easy to arrive at a conclusion that the angle between the slip plane and minor principal stress can be $\pi/4 + \overline{\phi}/2$ from Mohr theory. It then draws the conclusion that the minor principal stress must be horizontally acting on the right edge of the arch. At the left of the arch, the horizontal normal stress on the wall at depth z, σ_{ahw}, can be calculated by considering the force equilibrium in triangular element near the wall, as shown in the bottom of Figure 1. Being equivalent of horizontal normal stress on the wall, the lateral stress can be

$$\sigma_{ahw} = \sigma_1 \cos^2\theta + \sigma_3 \sin^2\theta, \tag{3}$$

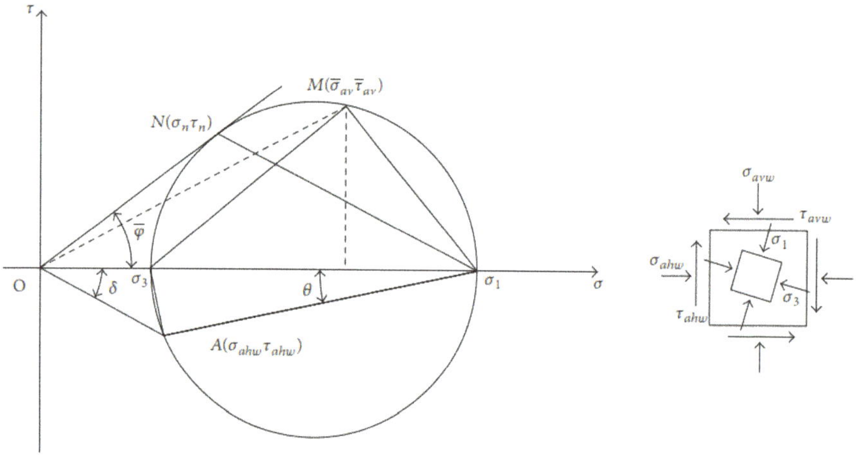

Figure 2: Mohr circle for earth stresses near the wall.

where θ describes the direction of the major principal stress with respect to horizontal near the wall. The conjugate vertical normal stress is expressed by

$$\sigma_{avw} = \sigma_1 \sin^2\theta + \sigma_3 \cos^2\theta. \tag{4}$$

Similarly, the horizontal normal stress at the random point along the arch with angle ψ to the horizontal can be expressed by

$$\sigma_{ah} = \sigma_1 \cos^2\psi + \sigma_3 \sin^2\psi. \tag{5}$$

The conjugate vertical normal stress is given by

$$\sigma_{av} = \sigma_1 \sin^2\psi + \sigma_3 \cos^2\psi. \tag{6}$$

Dividing (5) and (6) by σ_1 and substituting $\sigma_3/\sigma_1 = 1/N$ can derive

$$\frac{\sigma_{ah}}{\sigma_1} = \cos^2\psi + \frac{1}{N}\sin^2\psi,$$

$$\frac{\sigma_{av}}{\sigma_1} = \sin^2\psi + \frac{1}{N}\cos^2\psi. \tag{7}$$

Based on the theory of Mohr circle, as shown in Figure 2, the shear stress, τ_{ahw}, on the left edge of the arch is given by

$$\tau_{ahw} = \sigma_{ahw}\tan\delta = (\sigma_{ahw} - \sigma_3)\tan\theta. \tag{8}$$

Thus,

$$\tan\theta = \frac{\sigma_{ahw}\tan\delta}{\sigma_{ahw} - \sigma_3}. \tag{9}$$

Substitution of (3) into (9) yields a quadratic equation

$$\tan\theta = \frac{\left(\sigma_1\cos^2\theta + \sigma_3\sin^2\theta\right)\tan\delta}{\sigma_1\cos^2\theta - \sigma_3\cos^2\theta} = \frac{\left(N + \tan^2\theta\right)\tan\delta}{N-1}. \tag{10}$$

Solving (10) for tan θ gives

$$\tan\theta = \frac{(N-1) \pm \sqrt{(N-1)^2 - 4N\tan^2\delta}}{2\tan\delta}. \tag{11}$$

For fear of the error propagation, it is an appropriate methodology to replace tan θ by character k. Of the two values given by (11), the larger one dovetails nicely with the experiments by Guo and Zhou [28].

The lateral earth pressure coefficient can be defined by the ratio of the horizontal stress to the vertical stress. As for a slice with depth z from the surface of the backfill, the lateral earth pressure coefficient is bound up with the horizontal normal stress and the average vertical normal stress. The vertical normal stress acting on the unit arch, corresponding to the given slice, can be calculated as

$$dV = \sigma_1\left(\sin^2\psi + \frac{1}{N}\cos^2\psi\right)(R\sin\psi d\psi), \tag{12}$$

Where R is the radius of the arch, representing minor principal stress trajectory, and it can be given by

$$R = \frac{H - z}{1 - \sin\theta + \cos\theta\tan\alpha}. \tag{13}$$

The vertical component of average normal stress along the arch can be calculated by

$$\overline{\sigma}_{av} = \frac{\int_{\theta}^{\pi/2} \mathrm{d}V}{R\cos\theta} = \int_{\theta}^{\pi/2} \sigma_1 \left(\sin^2\psi + \frac{1}{N}\cos^2\psi\right)\frac{\sin\psi}{\cos\theta}\mathrm{d}\psi$$

$$= \sigma_1\left(1 - \frac{N-1}{3N}\cos^2\theta\right). \tag{14}$$

Dividing (1) by (14), it can be derived that

$$K = \frac{\sigma_1\cos^2\theta + \sigma_3\sin^2\theta}{\sigma_1\left(1 - ((N-1)/3N)\cos^2\theta\right)} = \frac{3N\cos^2\theta + 3\sin^2\theta}{3N - (N-1)\cos^2\theta}$$

$$= \frac{3N + 3k^2}{1 + 2N + 3Nk^2}. \tag{15}$$

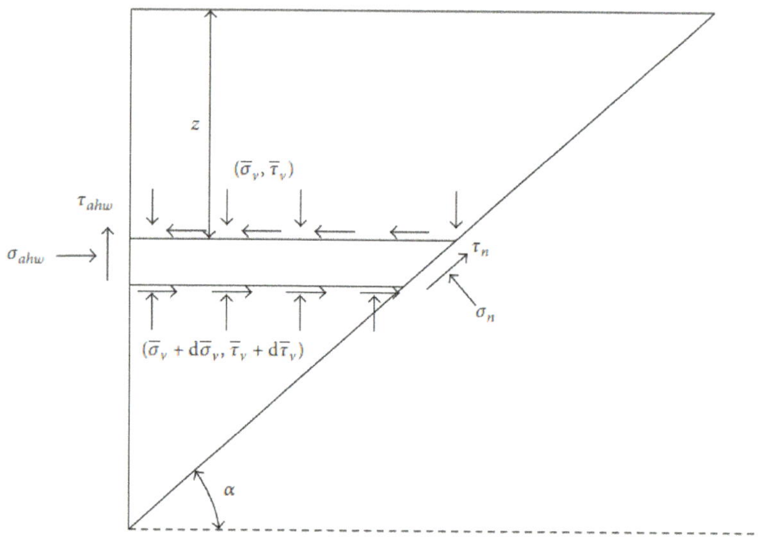

Figure 3: Free body diagram of differential slice.

From (15), it should lay great stress on the fact that the lateral earth pressure coefficient only has relation to the ratio of principal stresses ($N=\sigma_1/\sigma_3$) and the wall-soil friction angle (δ).

The shear stress, $\overline{\tau}_{av}$, corresponding to $\overline{\sigma}_{av}$ can be obtained in the following way. As shown in Mohr circle, it can be deduced from triangular diagram that

$$\overline{\tau}_{av}^{\,2} = (\sigma_1 - \overline{\sigma}_{av})(\overline{\sigma}_{av} - \sigma_3). \tag{16}$$

Substituting (14) in (16), it can be obtained that

$$\bar{\tau}_{av} = \frac{1}{3}\sigma_1 \cos\theta\left(1 - \frac{1}{N}\right)\sqrt{3 - \cos^2\theta}. \tag{17}$$

Rather, the coefficient ratios between $\bar{\tau}_{av}$ and $\bar{\sigma}_{av}$ are kept as the function of θ and can be exhibited as

$$K_{av} = \frac{(N-1)\cos\theta\sqrt{3-\cos^2\theta}}{3N-(N-1)\cos^2\theta}. \tag{18}$$

According to the trigonometric function, (18) can be transformed into

$$K_{av} = \frac{(N-1)\sqrt{2+3k^2}}{1+2N+3Nk^2}. \tag{19}$$

ANALYSIS OF EARTH PRESSURE ON THE RETAINING WALL OF RB

The differential slice technique is widely applied for the limit equilibrium condition where an active horizontal displacement of sliding wedge has developed in the soil mass behind a retaining wall. The technique has not been used for valid distribution of lateral earth pressures which is necessary in the design of walls in which the shearing resistance of the soil mass behind the wall is not completely mobilized according to rotation and earth arching effects. Based on the differential slice technique, the following is to discuss the earth pressure, resultant earth force, and its location of application on the retaining wall rotating about base considering arching effects. At all events, friction always serves itself as a force that opposes motion. Distinguished from differential slice technique for translation mode, it should take into account friction forces between slices for RB mode. Figure 3 shows local force equilibrium of differential slice at the depth z from the surface of the backfill behind the rotating wall. σ_{ahw} is the horizontal components acting by retaining wall; τ_{ahw} is the vertical components caused by friction between the wall and the soils; $\bar{\sigma}_v$ and $\bar{\tau}_v$ represent the mean vertical normal stress and the mean shear stress acting on the top, respectively; $\bar{\sigma}_v + d\bar{\sigma}_v$ and $\bar{\tau}_v + d\bar{\tau}_v$ represent the mean vertical normal stress and the mean shear stress acting on the bottom, respectively; σ_n and τ_n represent the normal stress and the shear stress acting on the slipping surface, respectively; α is the angle of slipping surface respecting the horizontal; dW is the weight of the differential slice; γ is the bulk density of the soil. The equilibrium of force for horizontal can be expressed as

$$\sigma_{ahw}\mathrm{d}z + \tau_n \frac{\mathrm{d}z}{\sin\alpha}\cos\alpha - \sigma_n \frac{\mathrm{d}z}{\sin\alpha}\sin\alpha - \overline{\tau}_v (H - z)\cot\alpha$$
$$+ (\overline{\tau}_v + \mathrm{d}\overline{\tau}_v)(H - z - \mathrm{d}z)\cot\alpha = 0. \tag{20}$$

After elided quadratic differential item, (20) is simplified:

$$\sigma_{ahw}\mathrm{d}z + \tau_n \cot\alpha\mathrm{d}z - \sigma_n\mathrm{d}z + (H - z)\cot\alpha\mathrm{d}\overline{\tau}_v - \overline{\tau}_v\cot\alpha\mathrm{d}z$$
$$= 0. \tag{21}$$

Summation of all forces in vertical acting on the differential slice yields

$$\mathrm{d}\overline{\sigma}_v (H - z)\cot\alpha + \tau_{ahw}\mathrm{d}z + \sigma_n\mathrm{d}z\cot\alpha + \tau_n\mathrm{d}z$$
$$- \overline{\sigma}_v\mathrm{d}z\cot\alpha - \mathrm{d}W = 0. \tag{22}$$

Because of the overt relationship between normal stresses and shear stresses along the wall and along the slipping surface, it is liable to reach the following expressions:

$$\sigma_{ahw} = K\overline{\sigma}_v,$$
$$\tau_n = \sigma_n \tan\phi,$$
$$\tau_{ahw} = \sigma_{ahw}\tan\delta,$$
$$\overline{\tau}_v = K_{av}\overline{\sigma}_v,$$
$$\mathrm{d}W = \gamma (H - z)\cot\alpha\mathrm{d}z. \tag{23}$$

Substituting (23) into (21) and (22), the yielded simultaneous equations could be reduced as

$$\frac{\mathrm{d}\overline{\sigma}_v}{\mathrm{d}z} = A\gamma - \frac{(1 - BK)\overline{\sigma}_v}{H - z}, \tag{24}$$

Where

$$A = \frac{(1 - \tan\phi\cot\alpha)}{C},$$
$$B = \frac{(1 + \tan\alpha\tan\phi + \tan\delta\tan\alpha - \tan\delta\tan\phi)}{C},$$
$$C = 1 + \cot\alpha (K_{av} - \tan\phi) + K_{av}\tan\phi. \tag{25}$$

Further, it should be emphasized that the stress integration in (24) over the height must be in accordance with initial condition of $\overline{\sigma}_v = 0$ at $z = 0$. Thus, $\overline{\sigma}_v$ can be expressed as

$$\overline{\sigma}_v = \frac{AH\gamma}{BK-2}\left(\left(1-\frac{z}{H}\right)-\left(1-\frac{z}{H}\right)^{BK-1}\right). \tag{26}$$

The horizontal lateral pressure can be got with a given lateral pressure coefficient

$$\sigma_{ahw} = K\overline{\sigma}_v = \frac{KA\gamma H\left((1-(z/H))-(1-(z/H))^{BK-1}\right)}{BK-2} \tag{27}$$

and then the horizontal component of the resultant force can be the integration of (27):

$$F_h = \int_0^H \frac{KA\gamma H\left((1-(z/H))-(1-(z/H))^{BK-1}\right)}{BK-2}\mathrm{d}z$$

$$= \frac{A\gamma H^2}{2BK}. \tag{28}$$

As shown in Figure 1, the total resultant thrust can be calculated as

$$F_T = \frac{F_h}{\cos\delta} = \frac{A\gamma H^2}{(4-2BK)\cos\delta}. \tag{29}$$

The application point of the total force is the focus in engineering practice. For the distribution of lateral earth pressure is nonlinear, the height of application of the earth force on the wall can be obtained by dividing the moment of the lateral earth pressure about the base by the lateral horizontal earth force. The moment, M, can be the integral of moment of slices about the base

$$M = \int_0^H \frac{KA\gamma H\left((1-(z/H))^{1-BK}-(1-(z/H))\right)}{BK}(H-z)\,\mathrm{d}z$$

$$= \frac{KA\gamma H^3}{3+3BK} \tag{30}$$

and the location of application of the total earth force is expressed as

$$l = \frac{M}{F_h} = \frac{2BKH}{3+3BK}. \tag{31}$$

RESULTS AND DISCUSSION

Results of observation from Fang and Ishibashi [13] model studies furnish the basis for the present discussions. The tests involved a 1 m high wall rotating about the base in a loose to medium dense sand ($\overline{\phi} = 33.4^\circ \rightarrow 40.4^\circ$ and $\gamma = 15.4$ kN/m^3). The limiting value of wall friction δ is always treated as a linear function of internal friction ϕ and expressed as $\delta = a\overline{\phi}$. With respect to two cases discussed here, it can conservatively replace δ by $\delta = 2\overline{\phi}/3$. Figure 4 plots the calculated values of the point of application of the active thrust with respect to the soil shear strength. The results indicate that the arching effects bring about higher location of active resultant forces than those without considering arching effects behind walls rotating about the base. The comparison also reveals that only small differences exist between the results with arching effects and the measured datum.

Figure 5 shows a comparison of calculated lateral pressures with Fang's measurements [13] for the case of RB in sand ($\overline{\phi}$ = 33.4°). Such comparison is also done with Rankine theory [2] and Choudhury theory [7] as shown in Figure 5. This case is viewed as quasi-active state for the shearing resistance at the base becoming fully mobilized [15]. The curve, plotted as the normalized mobilized lateral pressure in equilibrium state, $\sigma_V/\gamma H$, versus the normalized depth, z/H, clearly shows the nonlinear lateral earth pressure distribution. The trajectory agrees with the experimental data by Fang and Ishibashi and the calculated resultant force is about only 2.64% less than the active Rankine value but has some difference with Choudhury theory [7].

Prior to rotation, the stress state in the soil will correspond to the initial at-rest condition as naturally deposited sand.

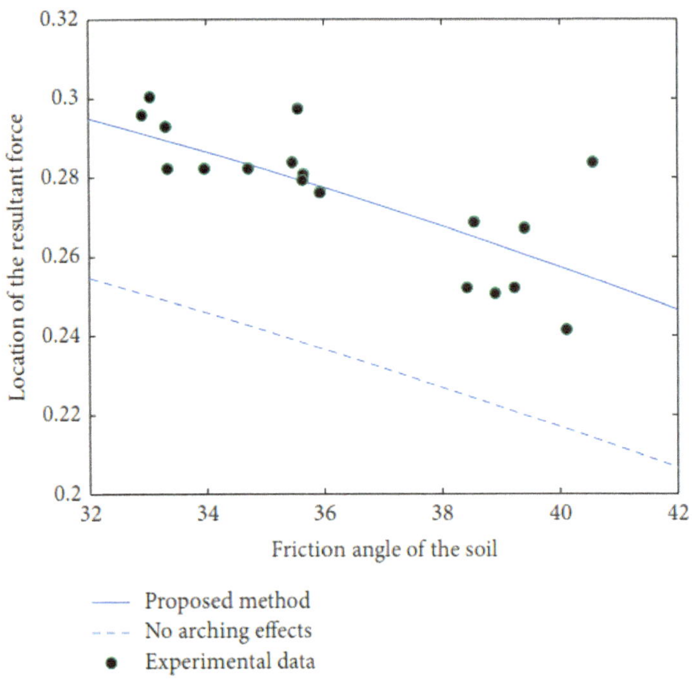

Figure 4: Effect of $\overline{\phi}$ on normalized height of application of thrust.

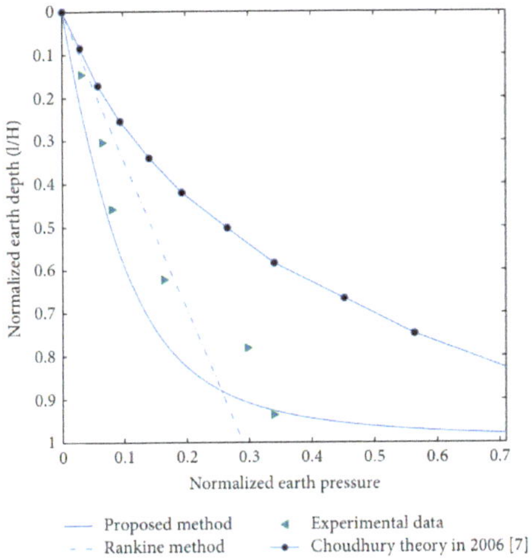

Figure 5: Distribution of normalized earth pressure with different methods for the case $\overline{\phi} = 33.4^\circ$.

In virtue of that mobilization of shearing resistance will propagate downwards as the angle of rotation increases [29]; it should select a proper initial angle of internal friction of the soil for at-rest condition. As method of Chang's model [15], the following case chooses $\phi_0 = 10.4°$ for $\overline{\phi} = 33.4^\circ$. Figure 6 shows a similar comparison between several analytical methods and experimental results for RB in conformity to partially mobilized sand ($\phi_0 = 10.4^\circ$). The comparisons exhibit that the present method fits the experiment datum better than Choudhury's theory [7]. It is also revealed that the calculated pressure on the wall is only slightly deviated from the experimental data and that the calculated resultant force is only 2.33% less than result of Rankine method.

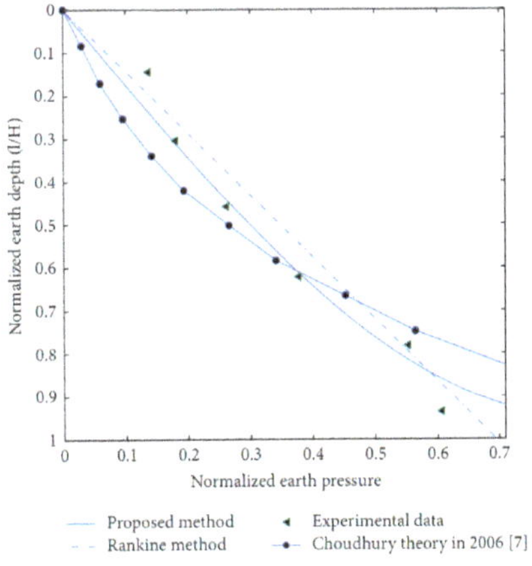

Figure 6: Distribution of normalized earth pressure with different methods for the case $\phi_0 = 10.4^\circ$.

CONCLUSION

An analytical method has been developed for the calculation of lateral earth pressure on the wall rotating about base considering arching effects in the backfill. The method uses differential slice technique and incorporates the deformation of stress distribution according to arching effects in the soil. Results calculated using lateral pressure coefficient that properly reflects the arching effects agree well with the model test data for backfilled walls rotating about the base.The following conclusions can also be drawn from aforementioned process.

(1) For a rotating wall about the base, an analytical method of shear stresses in slices is proposed in base of Mohr circle theory. It is simply exhibited as product of a scale coefficient with normal stress in horizontal slices.

(2) In the light of proposed method, the resultant lateral soil thrust exerted against the rotating wall about the base is less by only about 2.68% than the values estimated by the Rankine equation for fully mobilized case and by only about 2.35% for initial mobilized case. In contrast to single value of H/3 with Rankine theory, the variable point of application of the resultant force calculated by proposed method matches the experimental results which are higher than the value calculated by method without arching effects.

(3) As shown in comparison with Fang's experiment, the proposed method can reflect the nonlinear distribution of earth pressure on the wall and is identical to the datum of experiments.

Although the method developed here is primarily for cohesionless soil, the concept of the analysis and similar equations could be developed for soil with cohesion. Analysis for the passive case using this approach and the movement mode of the wall can be further discussed based on this work.

ACKNOWLEDGMENT

This work was supported by the Tianjin Research Program of Application Foundation and Advanced Technology under Grant no. 12JCZDJC28000. This support is greatly appreciated.

REFERENCES

1. C. A. Coulomb, Essai Sur une Application des règles de Maximis et Minimis à quelques Problèmes de Statique Relatifs à l'architecture, De l'Imprimerie Royale, 1776.

2. W. M. Rankine, On the Stability of Loose Earth, Philosophical Transactions of the Royal Society of London, 1857.

3. M. Khajehzadeh, M. R. Taha, A. El-Shafie, and M. Eslami, "Modified particle swarm optimization for optimum design of spread footing and retaining wall," Journal of Zhejiang University: Science A, vol. 12, no. 6, pp. 415–427, 2011. ·

4. R. Xu, "The GA-ANN method for determining calculation parameters for deep

excavation," Journal of Zhejiang University Science A, vol. 1, no. 4, pp. 408–413, 2000.

5. Y.-Z. Wang, "Distribution of earth pressure on a retaining wall," Geotechnique, vol. 50, no. 1, pp. 83–88, 2000. ·
6. K. Terzaghi, Theoretical Soil Mechanics, John Wiley & Sons, New York, NY, USA, 1943.
7. D. Choudhury and S. Chatterjee, "Displacement—based seismic active earth pressure on rigid retaining walls," Electronic Journal of Geotechnical Engineering, vol. 11, 2006.
8. D. Choudhury and S. Chatterjee, "Dynamic active earth pressure on retaining structures," Sadhana, vol. 31, no. 6, pp. 721–730, 2006.
9. D. Choudhury and S. S. Nimbalkar, "Pseudo-dynamic approach of seismic active earth pressure behind retaining wall," Geotechnical and Geological Engineering, vol. 24, no. 5, pp. 1103–1113, 2006. ·
10. K. S. S. Rao, S. Nayak, and D. Choudhury, "Determination of displacement-related passive earth pressure," Geotechnical Engineering, vol. 35, no. 2, pp. 79–85, 2004.
11. R. G. James and P. L. Bransby, "Experimental and theoretical investigations of a passive earth pressure problem," Géotechnique, vol. 20, no. 1, pp. 17–37, 1970.
12. M. A. Sherif, I. Ishibashi, and C. D. Lee, "Earth pressures against rigid retaining walls," Journal of the Geotechnical Engineering Division, vol. 108, no. 5, pp. 679–695, 1982.
13. Y.-S. Fang and I. Ishibashi, "Static earth pressures with various wall movements," Journal of Geotechnical Engineering, vol. 112, no. 3, pp. 317–333, 1986. ·
14. M. Matsuo, S. Kenmochi, and H. Yagi, "Experimental study on earth pressure of retaining wall by field tests," Soils and Foundations, vol. 18, no. 3, pp. 27–41, 1978. ·
15. M. F. Chang, "Lateral earth pressures behind rotating walls," Canadian Geotechnical Journal, vol. 34, no. 4, pp. 498–509, 1997. ·
16. F. Song and J.-M. Zhang, "Estimation of seismic earth pressures against rigid retaining structures with rotation mode," Open Civil Engineering Journal, vol. 5, no. 1, pp. 52–60, 2011. ·
17. B. Chevalier, G. Combe, and P. Villard, "Experimental and numerical study of the response of granular layer in the trap-door problem," in Powders and Grains, AIP Conference Proceedings, p. 649, July 2009.
18. B. Chevalier, G. Combe, and P. Villard, "Experimental and discrete element modeling studies of the trapdoor problem: influence of the macro-mechanical frictional parameters," Acta Geotechnica, vol. 7, no. 1, pp. 15–39, 2012. ·
19. K. Deb, "A mathematical model to study the soil arching effect in stone column-supported embankment resting on soft foundation soil," Applied Mathematical Modelling, vol. 34, no. 12, pp. 3871–3883, 2010.· View at MathSciNet

20. T. Eskişar, J. Otani, and J. Hironaka, "Visualization of soil arching on reinforced embankment with rigid pile foundation using X-ray CT," Geotextiles and Geomembranes, vol. 32, pp. 44–54, 2012. ·
21. K. Terzaghi, Soil Mechanics in Engineering Practice, John Wiley & Sons, New York, NY, USA, 1996.
22. R. L. Handy, "The arch in soil arching," Journal of Geotechnical Engineering, vol. 111, no. 3, pp. 302–318, 1985. ·
23. K. Harrop-Williams, "Arch in soil arching," Journal of Geotechnical Engineering, vol. 115, no. 3, pp. 415–419, 1989. ·
24. K. H. Paik and R. Salgado, "Estimation of active earth pressure against rigid retaining walls considering arching effects," Geotechnique, vol. 53, no. 7, pp. 643–653, 2003. ·
25. B. Jiang, H.-W. Ying, and K.-H. Xie, "Analysis on soil arching behind retaining wall," Journal of Zhejiang University (Engineering Science), vol. 39, no. 1, pp. 131–136, 2005.
26. V. Potts and L. Zdravkovic, "Finite element analysis of arching behaviour in soils," in Proceedings of the 12th International Conference on Computer Methods and Advances in Geomechanics, pp. 3642–3649, October 2008.
27. S. S. Nadukuru and R. L. Michalowski, "Arching in distribution of active load on retaining walls," Journal of Geotechnical and Geoenvironmental Engineering, vol. 138, no. 5, pp. 575–584, 2012. ··
28. P. Guo and S. Zhou, "Arch in granular materials as a free surface problem," International Journal for Numerical and Analytical Methods in Geomechanics, 2012. ·
29. H. Matsuzawa and H. Hazarika, "Analyses of active earth pressure against rigid retaining wall subjected to different modes of movement," Soils and Foundations, vol. 36, no. 3, pp. 51–65, 1996.

5
CHAPTER

SEISMIC STABILITY TIME-FREQUENCY ANALYSIS METHOD OF REINFORCED RETAINING WALL

Yang Changwei, Zhang Shixian, Zhang Jianjing and Bi Junwei

INTRODUCTION

In Wenchuan earthquake, only one type of destruction of reinforced retaining wall caused by earthquake ground motion results in the rigid-plastic complex reinforcement's fracture. This phenomenon fully shows that reinforced retaining wall is applicable in high intensity earthquake area [1]. The earthquake stability analysis methods for reinforced retaining wall are usually pseudostatic method, limit displacement method, and numerical method [2–7]. The pseudostatic method only considers the effect of PGA on the stability of retaining wall and ignores the duration and frequency of seismic wave; while duration and frequency are not considered in limit displacement method, the method is not convenient in engineering practice with poor precision. The progress of computer techniques promotes the numerical analysis methods development in practical engineering analysis. The methods contain the finite element method, boundary element method, the meshless method, and so forth [8, 9]. Among those analysis methods, the meshless method is suitable in resolving practical problems, and another problem as mesh distortion in the finite element method and boundary element method can be generally avoided. Now, there are dozens of meshless methods, and each method uses different weighted residuals and approximate functions [10], which might be the direction of development in numerical analysis methods. As for the design of the reinforced retaining wall, results are more precise than pseudostatic method, finite element method, and boundary element method. It can consider the effect of PGA likewise, duration, and frequency of seismic wave, and nonlinear behavior of backfill earth on the stability of retaining wall. However, the meshless method is hard for the engineers to learn and implement. Results of study also indicate that duration

Citation: Yang Changwei, Zhang Shixian, Zhang Jianjing, and Bi Junwei, Seismic Stability Time-Frequency Analysis Method of Reinforced Retaining Wall, http://dx.doi.org/10.1155/2015/178692.

and frequency of seismic wave have remarkable effect on the earthquake stability of reinforced retaining wall [11, 12]. It is thus important to strengthen time-frequency analysis method on the stability of reinforced retaining wall under earthquake excitation, which can be easily used in reinforced earth wall design and solves the problem of general design methods unable to consider the effect of frequency and duration of seismic wave. For example, this method is compatible in EXCEL or other simple procedures (see Figure 1).

Figure 1: Failure of reinforced retaining wall in Wenchuan earthquake.

According to the reports on retaining walls damage in the Wenchuan earthquake [13, 14], the SV-wave of which vibrating direction is perpendicular to the direction of propagation in the cross-section of retaining wall is the most destructive wave. Moreover, SV-wave is a complex nonstationary signal and its frequency alters gradually over time, restricting the practical usage of pseudostatic method and limited displacement method. To tackle the problem, this paper presents an analysis work based on the elastic wave theory [15] and summarizes the dynamic analysis model of reinforced retaining wall. For the case of expandability and nonexpandability lacing wires, the plane of the fracture being linear type is assumed and thus the first-order differential equation of the seismic active earth pressure is established by horizontal slices analysis method. After that, by use of Hilbert-Huang transform method [16], authors have proposed a time-frequency analysis method for internal antiseismic stability analysis of reinforced retaining wall under SV-wave excitation. The validation of the new method is illustrated and verified through shaking table test and numerical analysis. In an effort to study the advantages of the method, the seismic stability calculation of reinforced retaining wall is conducted and further compared with the relevant results from time-frequency analysis method, pseudostatic method, and numerical simulation method.

TIME-FREQUENCY ANALYSIS METHOD OF THE ANTISEISMIC STABILITY FOR REINFORCED RETAINING WALL

According to the above shortcomings of pseudostatic method and limited displacement method, this paper proposes a time-frequency analysis method, which can take fully into account the effect of three elements of the seismic wave on the antiseismic stability of reinforced retaining wall. Its derived process follows this train of thought: firstly, summarize the analysis model. The sliding mass of retaining wall is divided into several horizontal slices by slices method. Solve the force induced by seismic wave at infinitesimal body of any layer based on HHT and elastic wave theory, and then one can establish a mechanical balance equation of microelement. By such analogy, the governing equations of the whole sliding mass are deduced. Hence, the needed total tensile force and length of the lacing wires for the reinforced retaining wall are worked out. In essence, the inertial force is a simplified quantification [17] of seismic wave's propagation in the soil. Therefore, the horizontal seismic inertial force should be equivalent to the wave force in this paper.

Formula Derivation of Time-Frequency Analysis Method

Basic Assumptions

The time-frequency analysis method of the antiseismic stability for reinforced retaining wall depends on the following assumptions. (1) The location and shape of rupture planes in the backfill soil behind the retaining wall are very important to determine the cross-section and length of rods. In the specifications for design of highway reinforced earth engineering [18, 19], there are two kinds of simplified rupture planes for the geosynthetics with different modulus, and the rupture planes are shown in Figures 2 and 3. The first simplified rupture plane is linear type plane where dip angle is $45^\circ + \Phi/2$. This plane can divide the backfill soil into two sections, including active zone and passive zone. If the reinforcement is ductile, a number of field test results indicate that this rupture plane is close to the real failure plane. The second rupture plane is bend linear type plane, divided into two sections, which contains vertical section and inclined section. In the vertical section, the height is $h_1 = H/2$ and the distance from the retaining wall is b. In the inclined section, the height is $h_2 = H/2$ and the dip angle is $45^\circ + \Phi/2$. If the reinforcement is nonductile, field test results show that this rupture plane is close to the real failure plane. Therefore, according to the characteristics of rods, this paper simplifies the rupture plane to two kinds. (2) In the real reinforced earth retaining wall engineering, the backfill soil is placed in layers, while not every layer is proper continuous, homogeneous, and isotropic. As a theoretical study, this paper assumes that the backfill soil meets the above characteristics. A number of research results indicate that the above assumptions of backfill soil cannot produce bigger effect on the calculated results. (3) When the seismic wave spreads from the bottom of the retaining wall to the top, the internal material damping will dissipate major seismic energy, the seismic wave will produce waveform transformation, and the seismic wave will be transformed into surface wave, behaving in the elliptical polarization movement. The reflective seismic wave on the upper surface of soil carries a small amount of energy of downpropagation, which has been verified by the same results. In this paper, ignoring the influence of surface reflective seismic wave is reasonable.

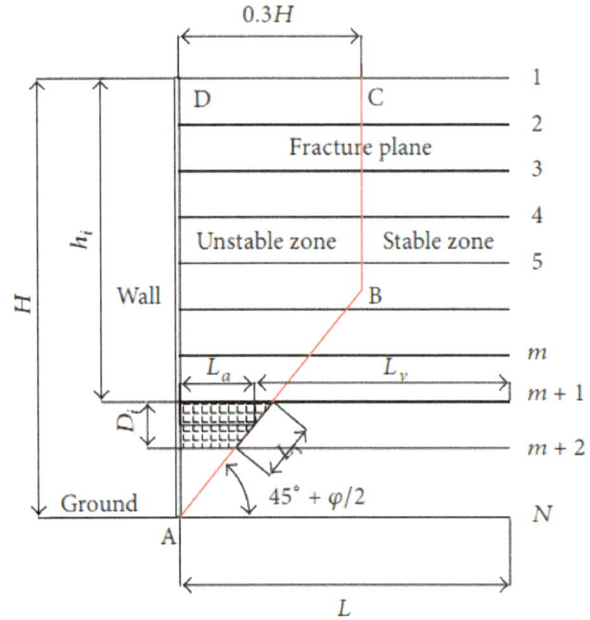

Figure 2: Nonductility reinforcement.

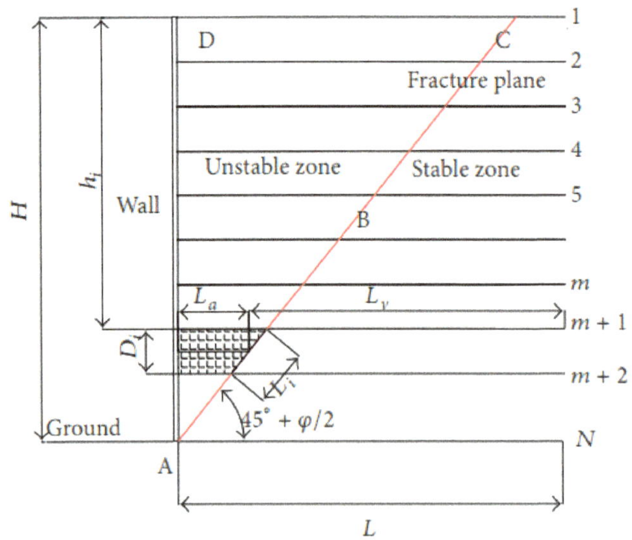

Figure 3: Reinforcement with ductility.

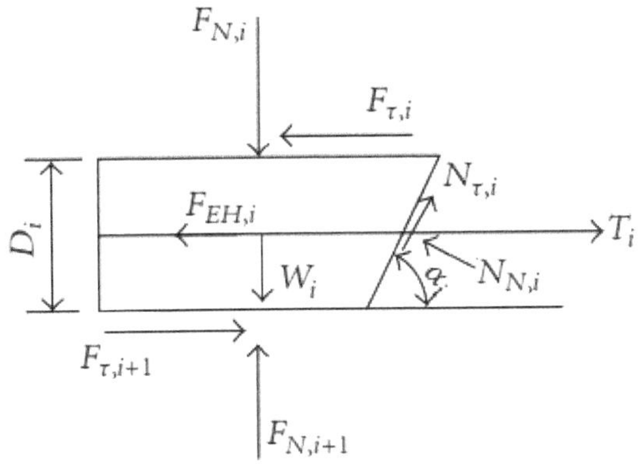

Figure 4: Analysis model of horizontal soil layer.

Generalized Model

For the cases of expandable and nonexpandable lacing wires, one can establish mechanical balance equations of the selective microelement by horizontal slice method, respectively. Figure 4 shows the parameters of unit-volume selected out as follows: the self-weight of the slice is W_i; the normal and shear interaction forces of the upper individual horizontal slice are $F_{N,}$, $F_{t,}$, respectively, and those of the lower slice are $F_{N,i-1}$, $F_{t,i-1}$; tensile force of the geosynthetic at layer i is T_i; the horizontal inertial force on the slice is $F_{EH,i}$; the angle between rupture plane and horizontal plane is αi. The free-body diagram of wave force on the horizontal slice is shown in Figure 5.

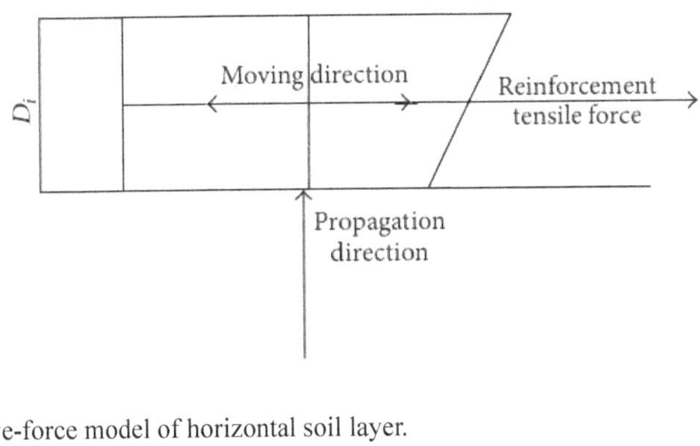

Figure 5: Wave-force model of horizontal soil layer.

Calculated Tension of Reinforcement.

In Figure 4, when the force of slice i reached a balance, the total of all vertical components is zero:

$$F_{N,i+1} - F_{N,i} - W_i + N_{t,i} \sin \alpha_i + N_{N,i} \cos \alpha_i = 0, \tag{1}$$

$$N_{t,i} = cl_i + N_{N,i} \tan\phi, \tag{2}$$

where c is the cohesion of the filling earth and φ is the internal frictional angle. Bring (2) into (3); the normal force of the horizontal slice $F_{N,}$ can be derived:

$$N_{N,i} = \frac{F_{N,i} - F_{N,i+1} + W_i - cl_i \sin\alpha_i}{\tan\varphi \sin\alpha_i + \cos\alpha_i}. \tag{3}$$

When the whole sliding mass reached a balance, the sum of all horizontal components is zero:

$$\sum_{i=1}^{n} T_i + \sum_{i=1}^{n} N_{t,i} \cos\alpha_i - \sum_{i=1}^{n} N_{t\cdot i} \sin\alpha_i - \sum_{i=1}^{n} F_{EH,i} = 0. \tag{4}$$

Bringing (2) and (3) into (4), we can get that

$$\sum_{i=1}^{n} T_i + \sum_{i=1}^{n} (cl_i + N_{N,i} \tan\phi) \cos\alpha_i - \sum_{i=1}^{n} \frac{F_{N,i} - F_{N,i+1} + W_i - cl_i \sin\alpha_i}{\tan\varphi \sin\alpha_i + \cos\alpha_i} \sin\alpha_i - \sum_{i=1}^{n} F_{EH,i} = 0. \tag{5}$$

The potential function of SV-wave can be formulated uniformly by elastic displacements [20], as follows:

$$u(z,t) = U(z)\, e^{i\omega t}, \tag{6}$$

where $U(z, t)$ is the displacement function of microunit i; $U(z)$ is the amplitude of the elastic displacement of microunit i; ω is the vibrational frequency of microunit i. Then, the inertial force which is generated by SV-wave of microunit i is as follows:

$$F_{EH,i} = \left| w_i \cdot \frac{\partial^2 u / \partial t^2}{g} \right| = \frac{W_i w^2 U(z)}{g}. \tag{7}$$

Hence, the needed total tensile force of the expandability and nonexpandability lacing wires in reinforced retaining wall can be obtained, respectively, by bringing (7) into (5), as follows:

$$\sum_{i=1}^{n} T_i = \rho V w^2 U(z) - \frac{cH\cos\alpha}{\sin\alpha} - \frac{[\cos\alpha\tan\varphi - \sin\alpha][F_{N,0} + W - cH]}{\tan\varphi\sin\alpha + \cos\alpha}, \tag{8}$$

$$\sum_{i=1}^{n} T_i = \rho V w^2 U(z) - 0.3cH + \frac{0.3(1 - 0.3\tan\alpha)\gamma H^2 - [c(1 - 0.3\tan\alpha)H]}{\tan\varphi} \times \left([\cos\alpha\tan\varphi - \sin\alpha]\left[F_{N,0} + 0.045\gamma H^2\tan\alpha\right] - 0.3cH\tan\alpha\right) \times (\tan\varphi\sin\alpha + \cos\alpha)^{-1}, \tag{9}$$

where $F_{N,0}$ is the overload at the top surface of the retaining wall; ω is the instantaneous frequency of SV-wave, which changes over time; ρ is the density of the filling earth behind retaining wall; V is the volume of the sliding mass ABCD; α is the angle between the rupture plane inclined part of retaining wall and the horizontal plane.

The needed total tensile force ($\sum_{i=1} T_i$) of the lacing wires in reinforced retaining wall can be transformed into the dimensionless value K, which is equal to the coefficient of soil pressure with the traditional design method for retaining wall. And then the formulation of K is that

$$K = \sum_{i=1}^{n} \frac{T_i}{\gamma H^2}. \tag{10}$$

The tensile force of lacing wires at layer i is T_i, as follows:

$$T_i = \gamma h_i D_i K. \tag{11}$$

Analysis Mentality of Time-Frequency Effect.

Timefrequency effect of seismic wave is mainly manifested in the elastic amplitude (z) and frequency ω of input seismic excitation. The analysis procedures of time-frequency effect proposed in this paper are as follows. (1) Decompose the seismic wave to several intrinsic mode functions (IMF) by the empirical mode decomposition (EMD) method. (2) Convert the multiple-channel multiple signals which consist of IMF to several single-channel single signals by conducting channel switch. (3) Solve instant frequency of each channel and plot the frequency time history curve. (4) The needed total tensile force of lacing wires can be obtained by bringing every IMF and its instant frequency into (8) and (9).

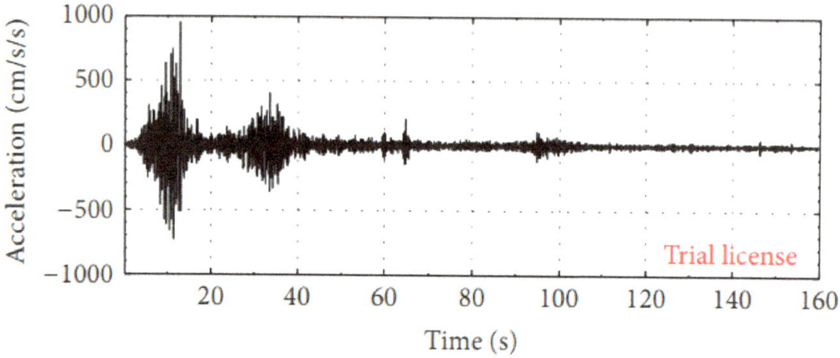

Figure 6: Time history of Wenchuan-Wolong wave.

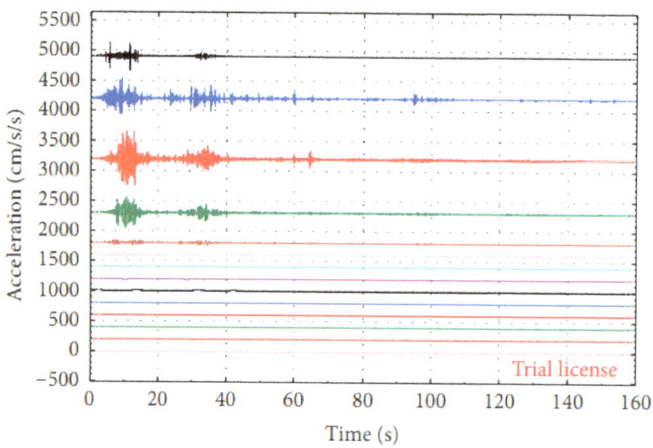

Figure 7: Acceleration time history of IMF.

From the above analysis, we know that time-frequency effect method of the antiseismic stability of reinforced retaining wall proposed in this paper can offset the drawbacks of pseudostatic method and limited displacement method. Moreover, it considers the effects of PGA, frequency, and duration primly for the stability of reinforced retaining wall under earthquake.

Application of Time-Frequency Effect for Seismic Wave.

In order to illustrate the application of time-frequency effect for seismic wave particularly, authors chose Wolong seismic waves input to carry out time-frequency analysis. (1) Input the Wolong seismic wave (Figure 6). (2) Disintegrate the seismic wave into multiple IMF by conducting EEMD method (Figure 7). (3) Solve the instant frequency of each IMF (Figure 8). Note that owing to sustenance over long periods of time for Wolong seismic wave, in Figures 7 and 8, the time history is only chosen from zero to eighty seconds, including peak ground acceleration.

Table 1: Comparison of tensile force of reinforcement between time-frequency and other calculated methods

PGA	Tensile force (kN) $\varphi = 20^\circ$			Tensile force (kN) $\varphi = 25^\circ$			Tensile force (kN) $\varphi = 30^\circ$		
	T-F	Shahgholi et al. [21]	Code	*T-F*	Shahgholi et al. [21]	Code	*T-F*	Shahgholi et al. [21]	Code
0.1	151	159	156	137	147	144	123	118	117
0.2	161	171	169	146	157	154	145	138	136
0.4	177	189	184	161	173	169	157	150	145

Note: PGA means peak ground acceleration; *T-F* means time-frequency method presented in this paper; Sha means method presented by Shahgholi et al., 2001 [21].

Table 2: Error analysis of tensile force of reinforcement between time-frequency and other calculated methods

PGA	Tensile force (%) $\varphi = 20^\circ$		Tensile force (%) $\varphi = 25^\circ$		Tensile force (%) $\varphi = 30^\circ$	
	Shahgholi et al. [21]	Code	Shahgholi et al. [21]	Code	Shahgholi et al. [21]	Code
0.1	5.30	3.14	7.30	4.76	−4.07	−5.08
0.2	6.21	4.68	7.53	5.10	−4.83	−6.52
0.4	6.78	3.70	7.45	4.62	−4.46	−8.00

Note: error = (the results of Shahgholi et al. [21] or Code-that of *T-F*) * 100%/the results of *T-F*.

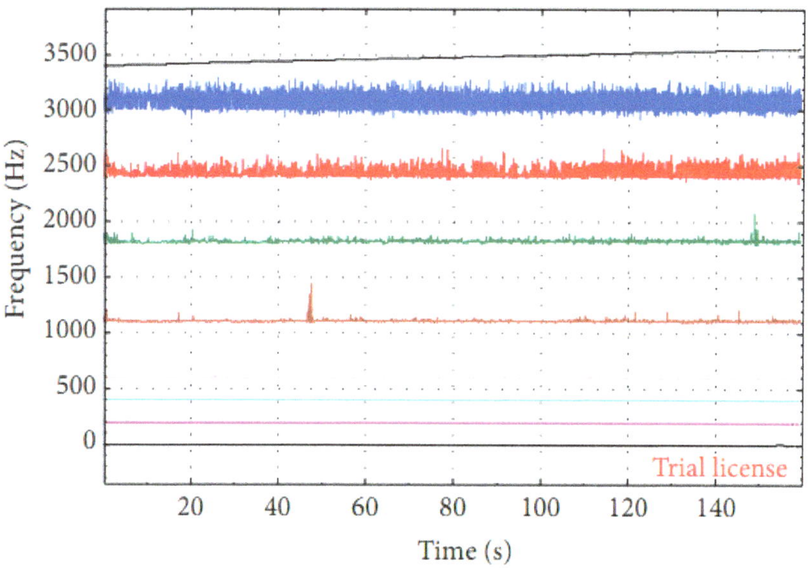

Figure 8: Frequency time history of IMF.

SOLUTION PROCEDURES OF TIME-FREQUENCY ANALYSIS METHOD FOR THE ANTISEISMIC STABILITY OF REINFORCED RETAINING WALL

The solution procedures of time-frequency analysis method for the antiseismic stability of reinforced retaining wall proposed in this paper are as follows.(1)Summarize the analysis model, based on the panel, filling earth, geosynthetics of reinforced retaining wall, and actual site condition.(2)Judge the place of fracture plane and bar spacing in accordance with the physical characteristic of soil and the type of reinforcement.(3) Determine the design response spectra according to the "Code for Seismic Design of Buildings," and then synthesize the acceleration time history by the method detailed in [22].(4)Generate displacement time history curve according to acceleration time history curve; then, with Hilbert-Huang transform, solve all the intrinsic mode functions of seismic wave and corresponding frequency time history curves.(5)Bring the results above into (8) or (9) to obtain the needed total tensile force of the lacing wires; moreover, in alliance with (10), acquire the tension of reinforcement in each layer.(6)Based on the above calculations and the position of rupture plane, we can get the requisite sectional area and the length of any reinforcement.

VERIFICATION OF TIME-FREQUENCY ANALYSIS METHOD FOR THE ANTISEISMIC STABILITY OF REINFORCED RETAINING WALL

In order to verify the correctness of time-frequency analysis method for the antiseismic stability of reinforced retaining wall, we consider the effect of the PGA, however, regardless of the frequency of seismic wave. For example, the unit weight of the backfill is 20 kN/m^3 and the height of retaining wall is 5 m, to arrange 18 layers of expandability lacing wires with constant spacing, and the antipull friction coefficient is 0.75; characteristic values of the shear strength parameters for the backfill are c=0 and $\varphi = 20^\circ$, 25°, 30°, respectively, PGA of the input sinusoidal wave is 0.1, 0.2, 0.4 (g), and the frequency is 15 HZ which is beyond the incidence of frequency about seismic wave by part 3. The results of the total tensile force are shown in Tables 1 and 2.

In general, the maximum error is 8.0%, and the minimum error is 3.14%, which can show that the results by timefrequency analysis method when ignoring the effect of frequency which are shown in Tables 1 and 2 are consistent with that by the current standard method and reference.

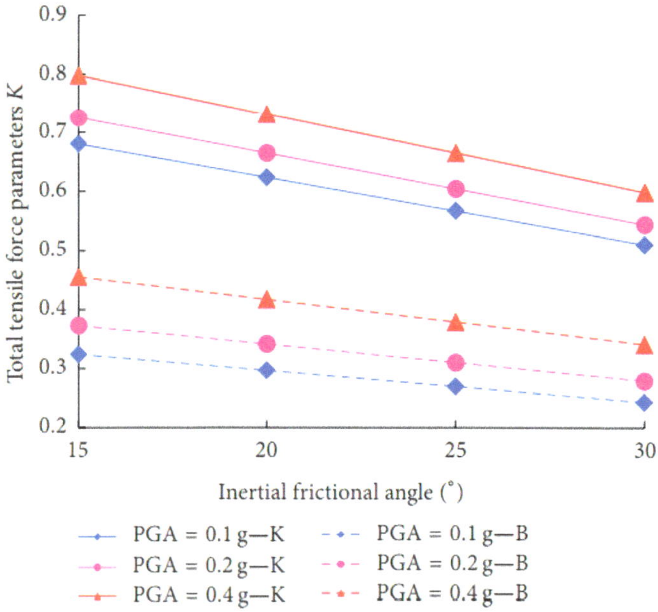

Figure 9: Relationship between total tensile force parameter *K* and internal friction angle of fill under frequency of 0.2 Hz.

Hence, the comparing simulation results proves the reliability and effectiveness of the new method proposed in this paper.

PARAMETRIC STUDY OF TIME-FREQUENCY ANALYSIS METHOD FOR THE ANTISEISMIC STABILITY OF REINFORCED RETAINING WALL

For the time-frequency analysis method of reinforced retaining wall, the internal frictional angle of the filling earth, the seismic intensity, and the frequency of the input earthquake wave in different types of geosynthetics have a predominant effect on the parameter *K*. For example, the unit weight of the backfill is 20 kN/m^3 and the height of retaining wall is 5 m; arrange 18 layers of reinforcements with a constant spacing and the antipull friction coefficient is 0.75; characteristic values of the shear strength parameters for the backfill are c=0 and φ = 15°, 20°, 25°, 30°, respectively, and PGA of the input sinusoidal wave is 0.1, 0.2, 0.4 (g) with the frequency change among 0.1, 0.2, 0.4, 0.8, 1.0, 2, 4, 5, 6, 8, and 10 (HZ). The values of the parameter *K* are shown in Figures 9, 10, 11, and 12, and as well *K* and *B* represent the expandable and nonexpandable lacing wires separately.

Analyzing Figure 9 to Figure 12 comprehensively, we can reach that the parameter *K* of the total tensile force of the lacing wires decreases with the increase of the internal frictional angle, and it may be because of the following reasons: the frictional resistance on the surface of lacing wires increases with the frictional angle, which improves the pulling resistance of lacing wires, and then the total tensile force reduces. However, the inertia force of backfill earth increases with the increase of the PGA, and then the total tensile force of lacing wires also increases significantly; the parameter *K* of the total tensile force of the expandability lacing wires is bigger than that of the nonexpandability lacing wires, and it may be because of the following reasons: if the reinforcement is ductility, the rupture plane is linear. On the contrary, the rupture is bending rupture plane that the reinforcement is nonductility. However, the wedge formed by the linear type plane is bigger than that formed by the bend linear type plane, and then the inertia force of the former is bigger than the latter, so the parameter *K* of the total tensile force of the expandability lacing wires is bigger than that of the nonexpandability lacing wires; the needed total tensile force of lacing wires is distributed in the shape of saddle and the force achieves the maximum value when the frequency of the input wave is 1 HZ. This phenomenon is because the instant frequency (1 HZ) of the input seismic wave is close to the natural frequency (1.16 HZ) of the retaining wall, causing the resonance effect to be enhanced; for the expandability and nonexpandability lacing wires, the calculations of *K* with the method that is proposed in this paper range from 1.0 to 1.25, while that value with normative methodologies between 1.02 and 1.35 (both inclusive), which indicate seismic safety reserve of retaining wall, may be lowered by formal methods for the design of antiseismic stability of the reinforcement retaining wall.

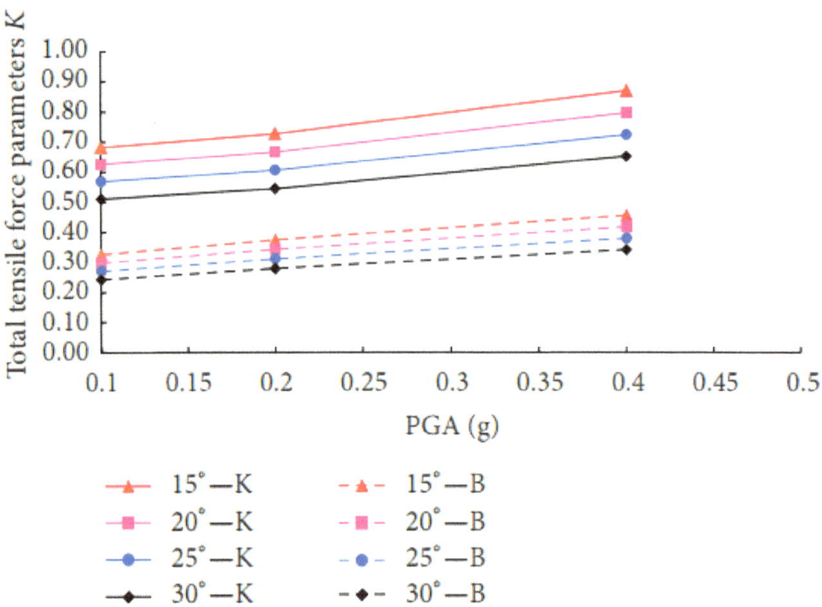

Figure 10: Relationship between total tensile force parameter *K* and PGA under frequency of 0.2 Hz.

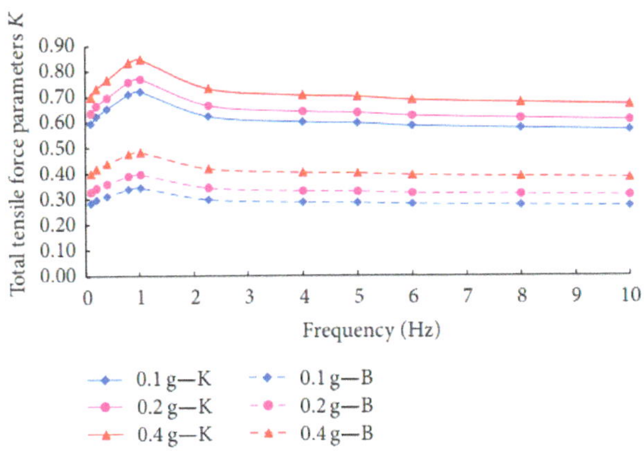

Figure 11: Relationship between total tensile force parameter K and frequency under internal friction angle of 20 degrees.

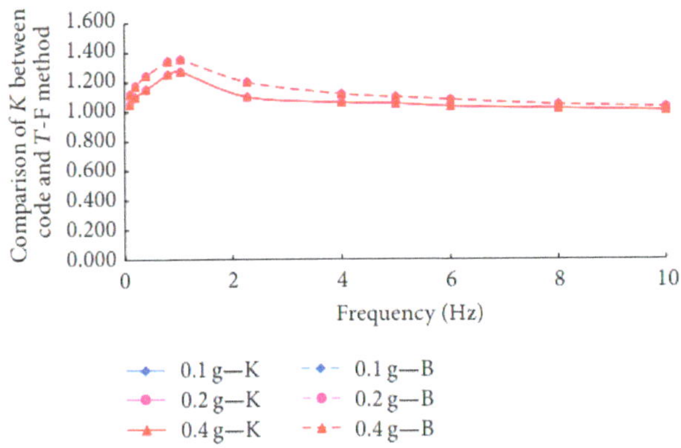

Figure 12: Relationship between different total tensile force parameter K and frequency under internal friction angle of 20 degrees.

From the above analysis, one can know that the type of geosynthetic, the internal frictional angle of the filling earth, the seismic intensity, and the frequency of the input earthquake wave have a predominant effect on the antiseismic stability of the reinforced retaining wall. Therefore, the effect of the above factors should be given due considerations in the earthquake-resistant design.

CONCLUSIONS

Based on the above analysis, the following conclusions can be concluded.

First, for the expandability and nonexpandability lacing wires, based on the elastic wave theory, we summarize the dynamic analysis model of the reinforced retaining wall and establish time-frequency analysis method of the antiseismic stability with HHT method by horizontal slices analysis method. And then the reasonability of this method is verified by the results of the other methods, for example, normative methodologies. The new method not only considers the effect of three factors of the seismic wave (PGA, frequency, and duration) on the internal antiseismic stability analysis of reinforced retaining wall and improves the seismic design accuracy greatly but also provides some valuable references for the seismic design of the other retaining structures.

Second, the type of geosynthetic, the internal frictional angle of the filling earth, the seismic intensity, and the frequency of the input earthquake wave have a predominant effect on the antiseismic stability of the reinforced retaining wall, which should be given due considerations in the earthquake-resistant design. Likewise, for the expandability and nonexpandability lacing wires, seismic safety reserve of retaining wall may be lowered by formal methods for the design of antiseismic stability of the reinforcement retaining wall when considering the effect of the PGA, while ignoring the frequency.

Third, the needed total tensile force of the lacing wires decreases with the increase of the internal frictional angle; on the contrary, it increases significantly with the increase of the PGA; the total tensile force of the expandability lacing wires is bigger than that of the nonexpandability lacing wires, and it is distributed in the shape of saddle and the force achieves the maximum value when the frequency of the input wave is the same as the natural frequency of reinforced retaining wall.

ACKNOWLEDGMENTS

This study is supported in part by NSF of China (Contract no. 51408510), Construction and Science and Technology Project of Ministry of Transport (2013 318800 020), Guangxi Science Foundation, the Program for Science and Technology of Henan Province in China (Grant no. 142300410200), and Science Foundation of Key Laboratory of High-speed Railway Engineering, Ministry of Education, School of Civil Engineering, Southwest Jiaotong University (Grant no. 2014-HRE-05).

REFERENCES

1. L. haiguang and Z. Depei, Design of the New Retaining Structure and Engineering Practice, Monographs, Communications Press, Beijing, China, 2004.
2. J. Jianqing, "Interior stability analysis of reinforced earth retaining wall under complicated dynamical loads," Central South Highway Engineering, vol. 32, no. 1, pp. 51–54, 2007.
3. J. Yingchun, "Internal stability calculation of reinforced retaining wall subjected to seismic load," Chinese Journal of Highway, vol. 10, no. 5, pp. 43–47, 1991.
4. H. B. Liu, "Internal stability analysis of segmental geosynthetic-reinforced soil retaining walls subjected to seismic loading," Chinese Journal of Geotechnical Engineering, vol. 30, no. 2, pp. 278–279, 2008.
5. M. M. El-Emam, Behaviour of reinforced soil walls under earthquake loading [Ph.D. thesis], Queen's University of Canada, Kingston, Canada, 2003.

6. Y. Youhai, "Stability analysis on the reinforced earth retaining wall under earthquake loadings," Journal of Lanzhou Railway University, vol. 21, no. 4, pp. 9–11, 2007.
7. P. Cui, Y.-Y. Zhu, Y.-S. Han, X.-Q. Chen, and J.-Q. Zhuang, "The 12 May Wenchuan earthquake-induced landslide lakes: distribution and preliminary risk evaluation," Landslides, vol. 6, no. 3, pp. 209–223, 2009.
8. L. W. Zhang, Z. X. Lei, K. M. Liew, and J. L. Yu, "Static and dynamic of carbon nanotube reinforced functionally graded cylindrical panels," Composite Structures, vol. 232, no. 111, pp. 205–212, 2014.
9. L. W. Zhang, P. Zhu, and K. M. Liew, "Thermal buckling of functionally graded plates using a local Kriging meshless method," Composite Structures, vol. 108, pp. 472–492, 2014.
10. X. Zhang, X.-H. Liu, K.-Z. Song, and M.-W. Lu, "Least-squares collocation meshless method,"International Journal for Numerical Methods in Engineering, vol. 51, no. 9, pp. 1089–1100, 2001.
11. C. B. Burke, Full-scale shaking table tests and finite element analysis of reinforced soil retaining walls [Ph.D. thesis], Columbia University, New York, NY, USA, 2004.
12. Y. Shengming, "Experimental study on reinforced soil retaining wall and analysis of deformation mechanism," in Monographs, pp. 9–11, Dalian University of Technology Press, Dalian, China, 2009.
13. C. Hua, The experimental study on mechanical characteristics of geogrid-reinforced soil retaining wall [Ph.D. thesis], China Academy of Railway Sciences, 2011.
14. J. Liang, Seismic response and dynamic stability analysis of reinforced retaining wall [Ph.D. thesis], Lanzhou University of Technology, Lanzhou, China, 2011.
15. X. Du, Theories and Methods of Wave Motion for Engineering, Science Press, Beijing, China, 2009.
16. N. E. Huang, Z. Shen, S. R. Long et al., "The empirical mode decomposition and the Hilbert spectrum for nonlinear and non-stationary time series analysis," The Royal Society of London A, vol. 454, no. 1971, pp. 903–995, 1998.
17. C. Yang, J. Zhang, and D. Zhou, "Research on time-frequency analysis method for seismic stability of rock slope subjected to SV wave," Chinese Journal of Rock Mechanics and Engineering, vol. 32, no. 3, pp. 483–491, 2013.
18. Ministry of Transport of the People's Republic of China, JTJ015-91 Specifications for Design of Highway Reinforced Earth Engineering, Standard, China Communications Press, Beijing, China, 1999.
19. U.S. Department of Transportation and Federal Highway Administration, FHWA2N HI2002043 Mechanically Stabilized Earth Walls and Reinforced Soil Slopes Design & Construction Guidelines, Standard, National Highway Institute Office of Bridge Technology, Washington, D.C., USA, 2002.

20. D. Xiuli, Theories and Methods of Wave Motion for Engineering, Monographs; Science Press, Beijing, China, 2008.
21. M. Shahgholi, A. Fakher, and C. J. F. P. Jones, "Horizontal slice method of analysis," Geotechnique, vol. 51, no. 10, pp. 881–885, 2001.
22. H. Yuxian and X. Xun, "Phase angle consideration in generating response spectrum-compatible ground motion," Earthquake Engineering & Engineering Vibration, vol. 6, no. 2, pp. 37–51, 1986.

6
CHAPTER

SEISMIC EARTH PRESSURES OF RETAINING WALL FROM LARGE SHAKING TABLE TESTS

Changwei Yang, Jian Jing Zhang, Qu Honglue, Bi Junwei, and Liu Fe-icheng

INTRODUCTION

The Wenchuan earthquake seriously damaged retaining structures of highway and railway in the disaster areas, which not only brings direct economic losses, but also delays the process of rescue and relief. To support the postdisaster reconstruction and complement the code for seismic design of retaining structures, Zhang et al. [1] carried out a number of field survey and analyses and mainly focused on the seismic damages of retaining structures. Results from the field survey show that the number of damaged gravity retaining walls is up to 97.1% of all damaged retaining walls. The result stipulates us to carry out seismic performance of gravity walls. Besides, estimation of the seismic earth pressure is an important topic of research for the safe design of retaining wall in the seismic zone. It is a common practice to the seismic accelerations in both horizontal and vertical directions in terms of equivalent static forces, called pseudostatic accelerations. Using the pseudostatic approach, several researchers have developed different methods to determine the seismic earth pressure on a rigid retaining wall due to earthquake loading starting from the pioneering works by Okabe [2] and Mononobe [3], commonly known as Mononobe-Okabe method [4] based on the pseudostatic approach, which gives the linear earth pressure distribution in a very approximate way irrespective of static and seismic conditions. Kumar [5] had determined the seismic passive earth pressure coefficients for sands using limit equilibrium method. Dewaiker and Halkude [6] have proposed a pseudostatic numerical analysis of seismic active and passive thrust on retaining wall, using Kotter's equation. Kumar and Chitikela obtained the seismic passive earth pressure coefficients using method of characteristics. Madhav and Kameswara Rao [7], Choudhury et al. [8],

Citation: Changwei Yang, Jian Jing Zhang, Qu Honglue, Bi Junwei, and Liu Feicheng, "Seismic Earth Pressures of Retaining Wall from Large Shaking Table Tests," Advances in Materials Science and Engineering, vol. 2015, Article ID 836503, 8 pages, 2015. doi:10.1155/2015/836503.

Choudhury [9], and Subba Rao and Choudhury [10] have adopted limit equilibrium for determining individually the seismic passive earth pressure coefficients corresponding to unit weight, surcharge, and cohesion components. Choudhury and Singh [11] have determined active earth pressure coefficients under static and seismic conditions using modified Culmann method. However, all the above methods are based on pseudostatic method of analysis, which does not consider the time effect of the applied earthquake load and the effect of shear and primary waves passing through the soil media. To overcome these drawbacks, the analytical method based on pseudodynamic approach as given by Steedman and Zeng [12] and modified by Choudhury and Nimbalker [13, 14] is used for the present analysis for calculation of seismic passive and active earth pressure.

Steedman and Zeng [12] considered in their analysis a vertical rigid retaining wall supporting a particular value of soil friction angle and a particular value of seismic horizontal acceleration (k_{hg}, where b is the acceleration due to gravity) only. Again they have considered the effect of horizontal seismic acceleration due to vertically propagating shear waves through the backfill behind retaining wall. In an improvement over this method, Choudhury and Nimbalkar [14] have incorporated effect of vertical seismic acceleration due to vertically propagating primary waves through the backfill soil. Again, they have studied the effect of various parameters such as wall friction angle, soil friction angle, shear wave velocity, primary wave velocity, and both the horizontal and vertical seismic accelerations on the seismic active earth pressure behind a rigid retaining wall by the pseudodynamic method. Choudhury and Nimbalker [13] have extended this modified work for estimation of seismic passive earth pressure.

In the pseudodynamic method, vertically propagating shear and primary waves through the backfill generate vibrations in horizontal and vertical directions, respectively. These horizontal and vertical vibrations correspond to horizontal and vertical time dependent seismic inertia forces, respectively. Time dependent nature of these seismic inertia forces is considered in the present analysis.

After the pseudodynamic method, the dynamic method which is finite element method and the shaking table test are used to the safe design of retaining wall in the seismic zone [15, 16].

In the procedure of the development of pseudostatic method, pseudodynamic method, and dynamic method, achievements have been obtained, but some problems have not been understood, such as the location of resultant earth pressure and relationships between earth pressure and ground shaking intensity. As a result, large shaking table tests are performed in order to effectively help us improve the understanding of characteristics of retaining structures in highway and railway engineering. The reason for performing the large shaking table tests is that the tests are the most direct research approach [17, 18].

SHAKING TABLE TESTS

Shaking Table Test

The tests were performed on a large shaking table with an advanced CNC system, data acquisition, and analysis system in the Nuclear Power Research and Design Institute of China. The shaking table, 6 m by 6 m, has 6 degrees of freedom; the maximum weight is 60 ton, the maximum horizontal displacement is ±150 mm, and the maximal vertical displacement is ±100 mm. The full-loaded horizontal and vertical maximum acceleration is 1.0 g and 0.8 g and the horizontal and vertical maximum acceleration for no loading is 3.0 g and 2.6 g with frequency ranges from 0.1 Hz to 80 Hz. And in the shaking table tests, a 128 channel BBM data acquisition system with a maximum referenced error $\leq 0.5\%$, is used to perform data acquisition, monitoring signal and analysis online synchronously, as shown in Figure 1.

Figure 1: Systems of CNC, data acquisition, and analysis.

Retaining Wall Model

In the shaking table tests, three different gravity retaining wall models are constructed in the tests: road shoulder retaining wall on rock foundation, road shoulder retaining wall on soil foundation, and embankment retaining wall on rock foundation.

The internal size of model boxes is 3.7 m × 1.5 m × 2.1 m and the model box is made of steel frame with steel plate on the bottom. In order to observe movement of the retaining wall and backfill, lateral walls of the steel frame are made of organic glass, and a type of absorbing material is set up on the lateral walls perpendicular to horizontal movement direction for reducing the reflection of seismic wave on the boundary. The height of the model wall is 1.6 m with a similitude ratio of 6 for modeling a wall with 9.6 m high and the backfill used in the test is granite sand, as shown in Figure 2. The similitude ratios for density and acceleration are 1, respectively, and similitude relationships of other parameters can be deduced based on Buckingham π theorem [19, 20]. The test model panorama is shown in Figure 3.

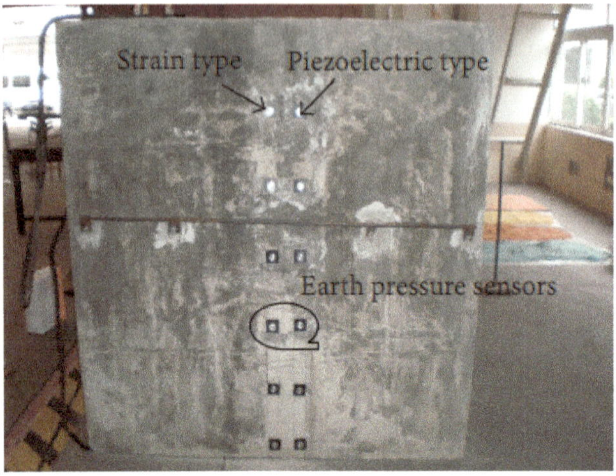

Figure 2: Gravity retaining wall model and earth pressure sensor installation diagram.

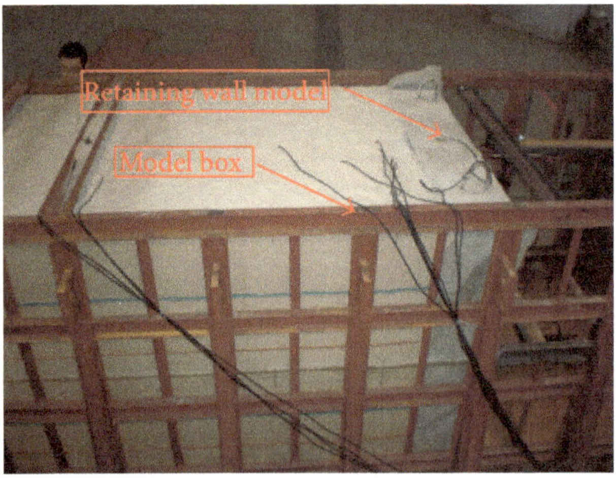

Figure 3: Test model panorama.

Data acquisition includes displacements, accelerations, and total and dynamic earth pressures. The sensor location arrangements are presented in Figure 4, and the sensor installation diagrams of earth pressures and displacements are shown in Figure 2 and Figure 5, respectively.

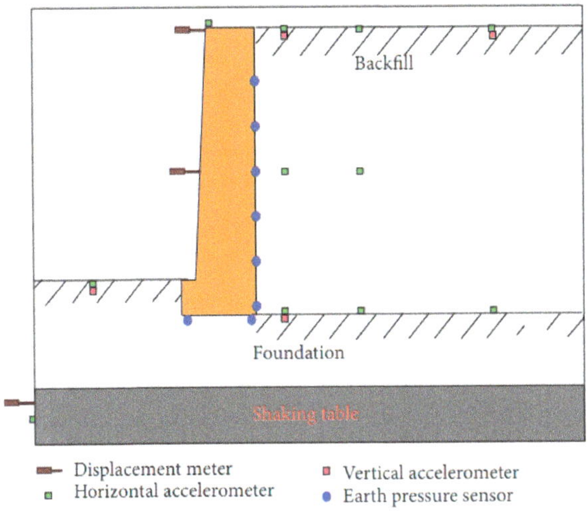

Figure 4: The instrument configuration.

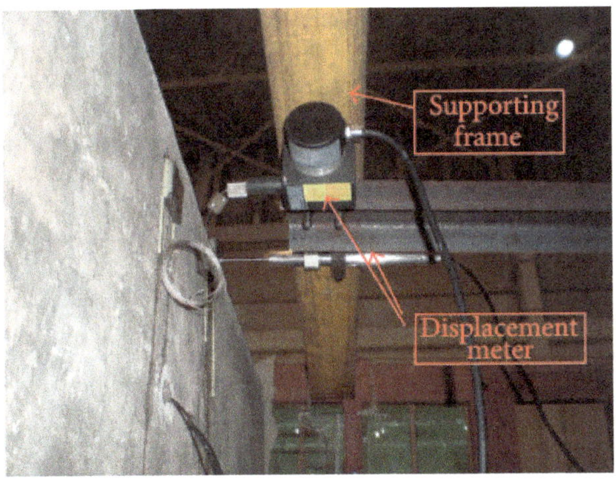

Figure 5: Displacement sensor installation diagram.

Loading Cases

Horizontal and vertical accelerations recorded at Wolong station in the Wenchuan earthquake are used to excite the model wall, which are scaled in the duration according to the similitude ratio used, as shown in Figure 6. Then the scaled seismic accelerations are scaled again on required amplitudes. The scaled seismic accelerations on duration and amplitude are used as input motions to excite the platform from the horizontal and vertical directions. The loading sequence in the test is white noise, PGA 0.1 g, PGA 0.2 g, PGA 0.4 g, PGA 0.7 g, and PGA 0.9 g.

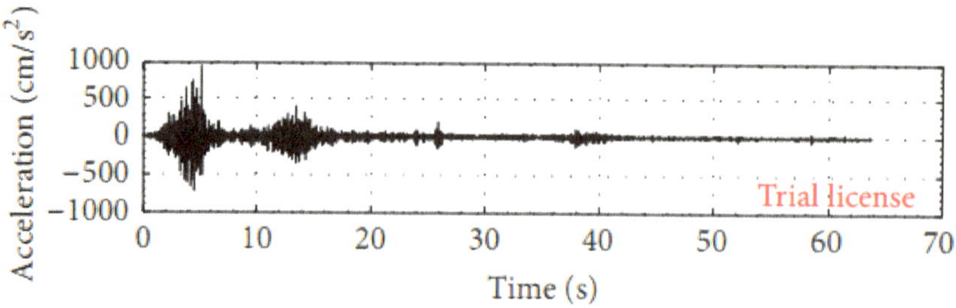

Figure 6: Scaled acceleration time history on duration and amplitude.

EXPERIMENTAL RESULT ANALYSES

Earth Pressure Distribution on Backwall

The measured distributions of total and dynamic earth pressures, as shown in Figures 7 and 8, present that the distributions are nonlinear along the wall height and the amplitude of the earth pressures are proportional to the excited seismic peak accelerations [21, 22]. The dynamic earth pressures are comparatively small for PGA less than 0.4 g but increase obviously for PGA larger than and equal to 0.4 g. Besides, the rate of earth pressure intensity change is related to geometric position, comparatively smaller above 2/3 H from toe of the wall but larger below.

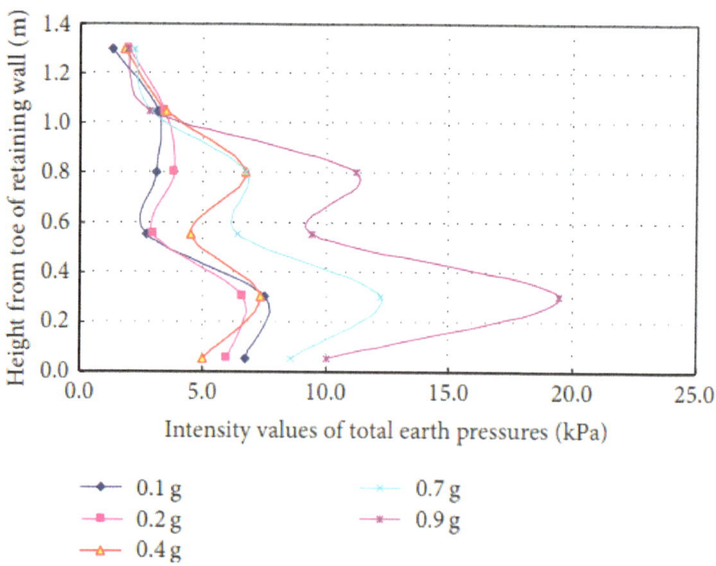

Figure 7: The distribution of total earth pressure.

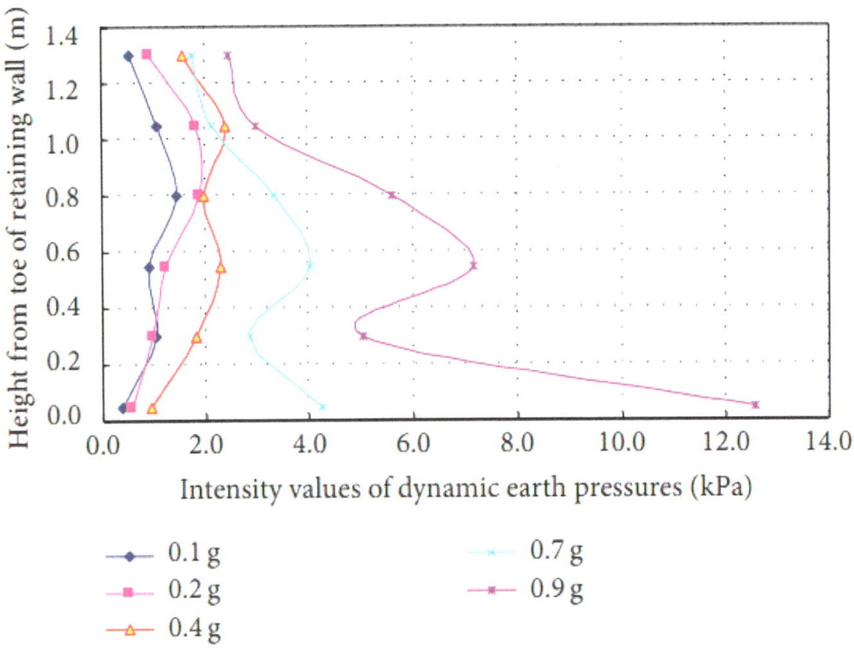

Figure 8: The distribution of dynamic earth pressure.

Resultant Earth Pressures on Backwall

Resultant dynamic and total earth pressures are proportional to the excited seismic peak acceleration and increase more obviously with the increase of excited seismic peak accelerations [23], as shown in Figures 9 and10. The foundation condition of retaining wall is an important influence factor to resultant total and dynamic earth pressure. Compared with retaining walls constructed on rock foundation, the dynamic and total earth pressures of retaining walls on soil foundation are larger. The influence of foundation condition is not obvious when the excited seismic coefficient is low, and the difference of total earth pressure is less than 15% for PGA less than or equal to 0.3 g but respectively, 20.56%, 52.84%, and 75.02% for PGA equal to 0.4 g, 0.7 g, and 0.9 g. It is clear that the impact of foundation condition becomes greater with the increase of excited seismic peak accelerations. As a result, the influence of foundation on a seismic performance of retaining wall should be considered according to the regions where the retaining wall locates. Comparing to the regions of seismic intensities below 8 degrees if a retaining wall locates, the influence of foundation in the regions with seismic intensities above 8 degrees should be carefully concerned [24].

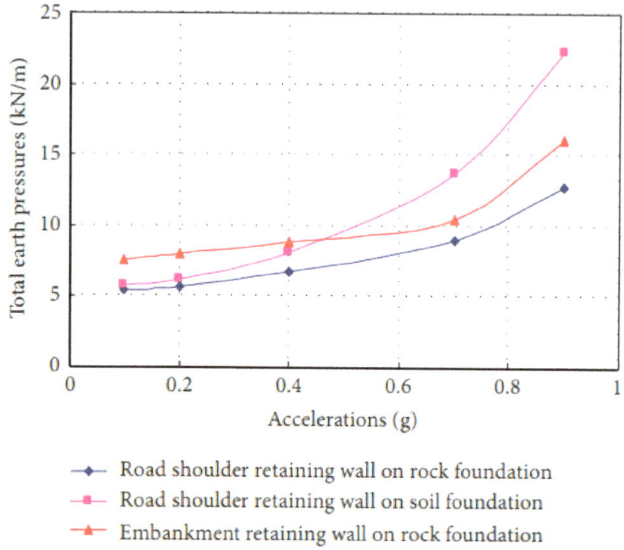

Figure 9: Resultant total earth pressures.

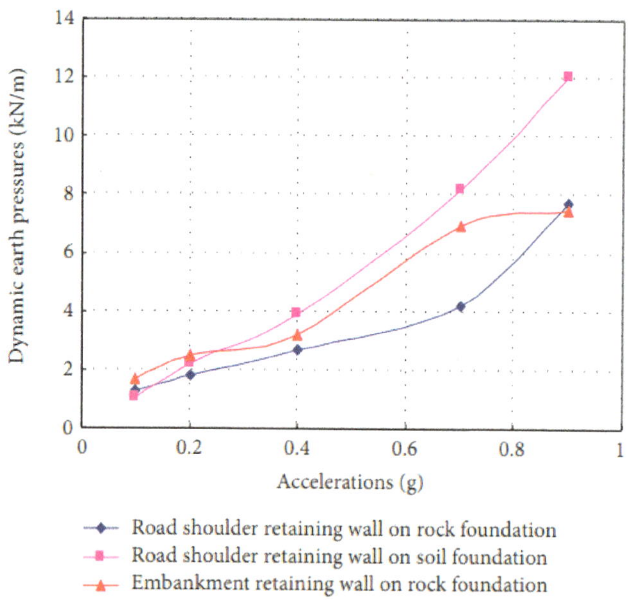

Figure 10: Resultant dynamic earth pressures.

In addition, the values of total and dynamic earth pressures are relevant to backfill form. The total and dynamic earth pressures of an embankment retaining wall are larger than those of shoulder retaining wall. The differences from the forms of backfill reduced with the increase of excited seismic peak acceleration but should not be ignored, as the embankment earth pressures are still larger.

Location of Resultant Earth Pressure on Backwall

The measured locations of resultant earth pressures (h/H) are summarized in Tables 1, 2, and 3. Obviously, when the seismic coefficient is low, measured locations of resultant total and dynamic earth pressure are above 0.33 H from the toe of wall. With the increase of seismic coefficient, the locations become close to 0.33 H, which is provided by M-O formula [2, 3]. Figure 11 presents the results of the locations of resultant total earth pressures (h/H) with excited seismic coefficients of three different test models, and Figure 12 presents the results of the locations of resultant dynamic earth pressures (h/H) with excited seismic coefficients of three different test models.

Table 1: Location of resultant earth pressure of road shoulder retaining wall on rock foundation (h/H)

Seismic coefficient	Location of resultant total earth pressure	Location of resultant dynamic earth pressure
0.1	0.34	0.46
0.2	0.38	0.49
0.4	0.39	0.48
0.7	0.34	0.4
0.9	0.33	0.35

* *H*: height of wall; *h*: height of location of resultant earth pressure from toe of wall.

Table 2: Location of resultant earth pressure of embankment retaining wall on rock foundation (h/H)

Seismic coefficient	Location of resultant total earth pressure	Location of resultant dynamic earth pressure
0.1	0.44	0.38
0.2	0.44	0.42
0.4	0.39	0.38
0.7	0.35	0.33
0.9	0.32	0.36

Table 3: Location of resultant earth pressure of road shoulder retaining wall on soil foundation (h/H)

Seismic coefficient	Location of resultant total earth pressure	Location of resultant dynamic earth pressure
0.1	0.63	0.77
0.2	0.61	0.76
0.4	0.55	0.64
0.7	0.48	0.46
0.9	0.4	0.46

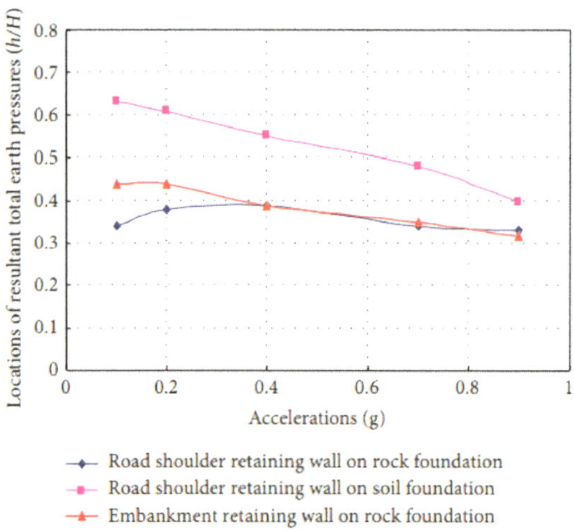

Figure 11: Locations of resultant total earth pressures (h/H).

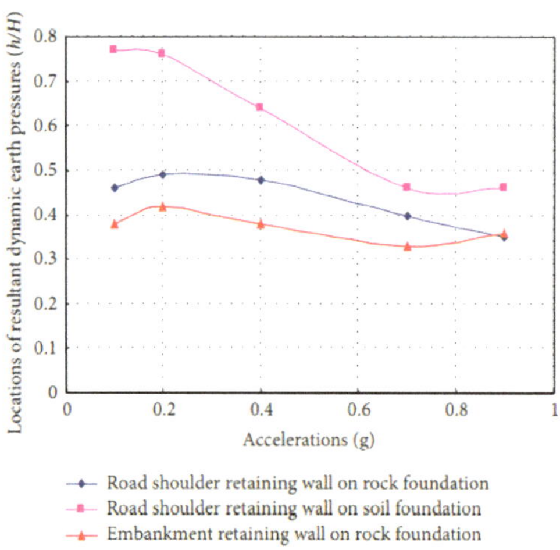

Figure 12: Locations of resultant dynamic earth pressures (h/H).

From Figures 11 and 12, we can see that, for a road shoulder retaining wall, the locations of resultant total earth pressure are lower than those of resultant dynamic earth pressure, and the difference gradually reduces with the increase of seismic coefficient [25]. Table 1 shows that the locations are between 0.33 H and 0.39 H for total earth pressures, while they are between 0.35 H and 0.49 H for dynamic earth pressures. For embankment retaining wall, the locations of resultant total and dynamic earth pressures are lower when compared with those for the road shoulder retaining wall, and when seismic coefficient is lower, the locations of resultant total earth pressure are higher than those of resultant dynamic earth pressure. With the increase of seismic coefficient, the locations of resultant total and dynamic earth pressure become close [26]. Table 2 shows that the locations are between 0.32 H and 0.44 H for total earth pressures, while they are between 0.33 H and 0.42 H for dynamic earth pressures.

The foundation conditions of retaining wall have a strong impact on the locations of resultant earth pressure. Normally, the softer the foundation is, the higher the locations are. For the shoulder retaining wall constructed on soil foundation, the resultant locations are between 0.4 H and 0.63 H for total earth pressures, while they are between 0.46 H and 0.77 H for dynamic earth pressures, as shown in Table 3.

Comparison of Measured Earth Pressures and Theoretical Results by Codes

The diagrams of comparison between measured earth pressures and theoretical results by codes on rock and soil foundation, respectively, are shown in Figures 13 and 14.

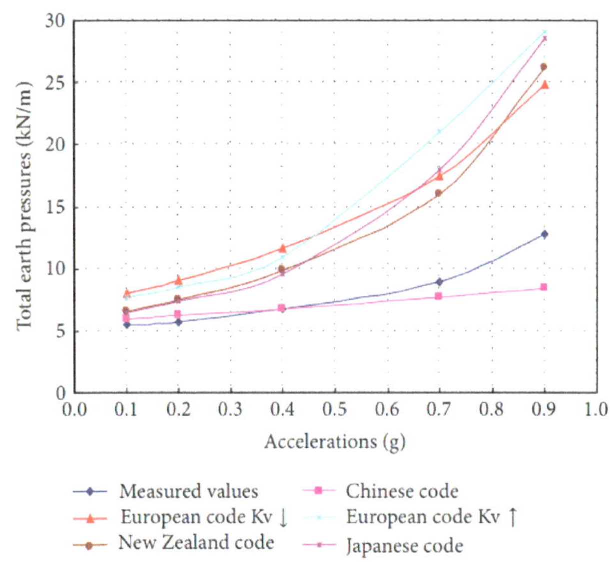

Figure 13: Comparison of measured earth pressures and theoretical results by codes for rock foundation.

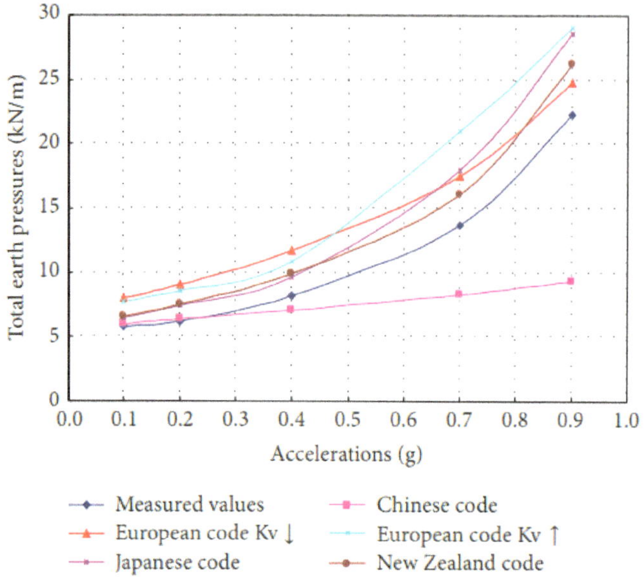

Figure 14: Comparison of measured earth pressures and theoretical results by codes for soil foundation.

Figures 13 and 14 show that, for rock foundation, the results by Chinese code are close to the measured earth pressures in the regions of seismic intensities below 8 degrees and smaller in the regions of seismic intensities above 8 degrees, while the results from European, Japanese, and New Zealand codes are conservative, contrast to Chinese code.

For soil foundation, the theoretical method by Chinese code for design of aseismic retaining wall is reasonable in the regions of seismic intensities below 8 degrees but is dangerous in the regions of seismic intensities above 8 degrees, as the theoretical results are obviously smaller than measured. Nevertheless, the other theoretical results by the European, Japanese, and New Zealand codes are slightly larger than the measured, and the difference between the codes is slight with seismic acceleration. In contrast to the Chinese code, these foreign codes are comparatively reasonable, and the results from the New Zealand code are closest to the measured ones.

Comparison of Measured Earth Pressures and Theoretical Results by Pseudostatic Method (Mononobe-Okabe Method) and Pseudodynamic Method [27]

In order to make a comparison of measured earth pressures and theoretical results by pseudostatic method and pseudodynamic method, this paper uses the same material parameters, earthquake wave, and the analysis model with the shaking table test for the rock foundation. The results are shown in Figures 15 and 17.

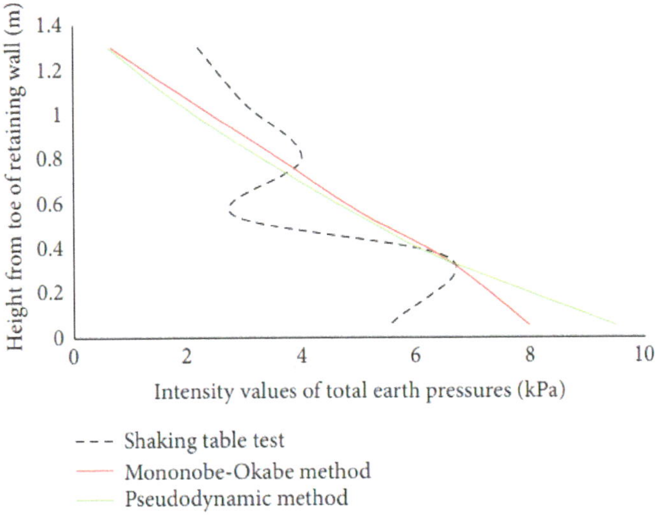

Figure 15: Comparison of measured earth pressures and theoretical results by pseudostatic method and pseudodynamic method for rock foundation (PGA of the input seismic wave is 0.4 g).

Figures 15, 16, and 17 show that, for the measured earth pressures, the results for the pseudostatic method and the pseudodynamic method are closed to shaking table test; for the total earth pressures and the location of resultant total earth pressures, the results for the pseudodynamic method are obviously closer to the shaking table test than those for pseudostatic method.

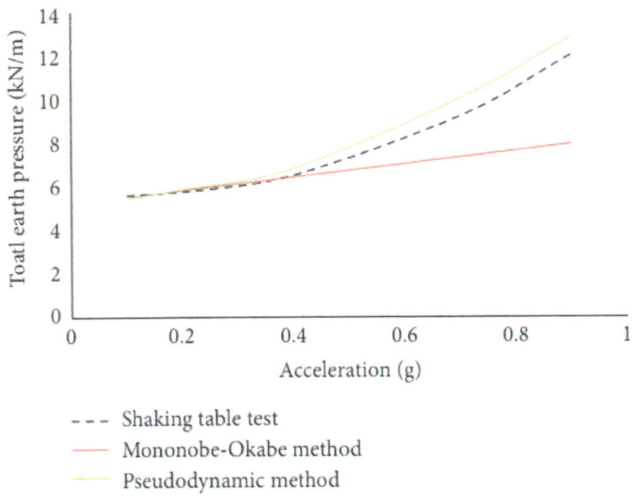

Figure 16: Comparison of measured total earth pressures and theoretical results by pseudostatic method and pseudodynamic method for rock foundation.

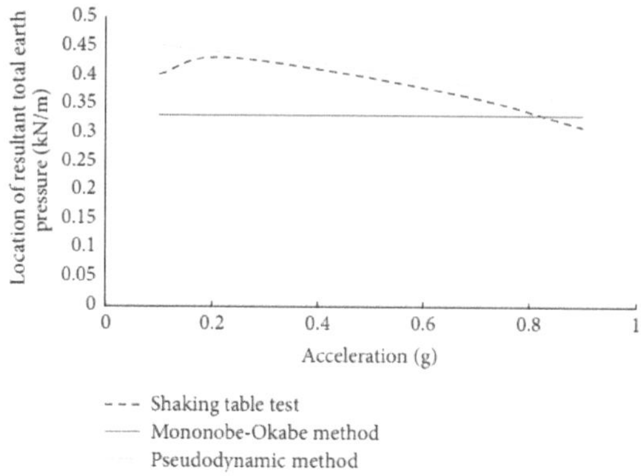

Figure 17: Comparison of locations of resultant total earth pressures (h/H) and theoretical results by pseudostatic method and pseudodynamic method for rock foundation.

Therefore, the results of pseudodynamic method are consistent with those of the shaking table test, but the results of pseudostatic method are smaller than measured.

CONCLUSIONS

(1) The distribution of total and dynamic earth pressure is nonlinear along the backwall. The dynamic earth pressures increase obviously for PGA larger than and equal to 0.4 g. Besides, the earth pressure intensity change speeds are related to geometric position, comparatively smaller above 2/3 H from toe of the wall but larger below.

(2) Resultant dynamic and total earth pressures grow more obviously along with the increase of seismic accelerations.

Comparing to rock foundation retaining wall, the dynamic and total earth pressures of soil foundation are larger. The foundation influence to design of aseismic retaining wall should be considered according to the regions and in the regions of seismic intensities above 8 degrees it should be carefully concerned.

The values of total and dynamic earth pressures are relevant to backfill form. The earth pressures of retaining wall on embankment are larger than on road shoulder. The influence from the form of backfill becomes less obvious along with seismic acceleration increasing but should not be ignored, as the numerical values are totally greater.

(3) When the seismic coefficient is low, measured locations of resultant total and dynamic earth pressure are above 0.33 H from the toe of wall, which is provided by M-O formula. With the rising of seismic coefficient, the locations become close to 0.33 H.

For retaining wall on road shoulder, the locations of resultant total earth pressure are lower than those of resultant dynamic earth pressure, and the gaps become closer with seismic coefficient rising. For retaining wall on embankment, the locations of resultant total and dynamic earth pressure are lower comparing to the retaining wall on road shoulder, and when the seismic coefficient is low, the locations of resultant total earth pressure are higher than those of resultant dynamic earth pressure. With the rising of seismic coefficient, the locations of resultant total and dynamic earth pressure become close.

The foundation condition has a strong impact on the locations of resultant earth pressure. At the same time, the softer, the locations higher.

(4) For rock foundation, the theoretical method for design of aseismic retaining wall provided by Chinese code is reasonable in the regions of seismic intensities below 8 degrees and smaller in the regions of seismic intensities above 8 degrees, while the results from those compared codes are too conservative.

For soil foundation, the method by Chinese code is reasonable in the regions of seismic intensities below 8 degrees but is dangerous in the regions of seismic intensities above 8 degrees. In contrast to the Chinese code, these foreign codes are comparatively reasonable, and the results from the New Zealand code are closest to the measured ones.

(5) The results of pseudodynamic are consistent with those of the shaking table test, but the results of pseudostatic method are smaller than measured.

ACKNOWLEDGMENTS

The authors wish to thank Dr. Hongsheng Ma and Bo Xiang for providing help in their field trip and survey. This study is supported in part by NSF of China (Contract no. 51408510), Guangxi Science Foundation, and the Program for Science and Technology.

REFERENCES

1. J. J. Zhang, J. Feng, S. G. Xiao, and C. Q. Liu, "Discussions on two key technical problems for seismic design of retaining structures," Journal of Southwest Jiaotong University, vol. 44, no. 3, pp. 321–326, 2009.

2. N. Okabe, "General theory on earth pressure and seismic stability of retaining wall and dam," Journal of Japan Society of Civil Engineers, vol. 10, no. 6, pp. 1277–1323, 1924.

3. N. Mononobe, "Considerations on vertical earthquake motion revelant vibration problems," Journal of Japan Society of Civil Engineerings, vol. 10, no. 5, pp. 1063–1094, 1924.

4. S. L. Kramer, Geotechnical Earthquake Engineering, Prentice Hall, New Jersey, NJ, USA, 1996.

5. J. Kumar, "Seismic passive earth pressure coefficients for sands," Canadian Geotechnical Journal, vol. 38, no. 4, pp. 876–881, 2001.

6. D. M. Dewaikar and S. A. Halkude, "Seismic passive/active thrust on retaining wall-point of application,"Soils and Foundations, vol. 42, no. 1, pp. 9–15, 2002.

7. M. R. Madhav and N. S. V. Kameswara Rao, "Earth pressure under seismic conditions," Soils and Foundations, vol. 9, no. 4, pp. 33–47, 1969.

8. D. Choudhury, K. S. SubbaRao, and S. Ghosh, "Passive earth pressures distribution under condition," inProceedings of the 15th International conference of Engineering Mechanics Division, ASCE, New York, NY, USA, June 2002.

9. D. Choudhury, "Seismic passive resisteance at soil-wall interface," in Proceedings of the 17th ASCE Engineering Mechanics Conference, University of Delaware, Netwark, Del, USA, 2004.

10. K. S. Subba Rao and D. Choudhury, "Seismic passive earth pressures in soils," Journal of Geotechnical and Geoenvironmental Engineering, vol. 131, no. 1, pp. 131–135, 2005.

11. D. Choudhury and S. Singh, "New approach for estimation of static and seismic active earth pressure,"Geotechnical & Geological Engineering, vol. 24, no. 1, pp. 117–127, 2006.

12. R. S. Steedman and X. Zeng, "Centrifuge modeling of the effects of earthquakes on free cantilever walls," in Centrifuge, vol. 91, pp. 425–430, Balkema, Rotterdam, The Netherlands, 1991.

13. D. Choudhury and S. Nimbalkar, "Seismic passive resistance by pseudo-dynamic method," Geotechnique, vol. 55, no. 9, pp. 699–702, 2005.

14. D. Choudhury and S. S. Nimbalkar, "Pseudo-dynamic approach of seismic active earth pressure behind retaining wall," Geotechnical and Geological Engineering, vol. 24, no. 5, pp. 1103–1113, 2006.

15. T. Xia, W. Hua, and Z. Wang, "Analysis of seismic active earth pressure of cohesive soil behind inclined retaining wall," World Earthquake Engineering, vol. 26, no. 1, pp. 315–321, 2010.

16. W. Liyan and L. Hanlong, "Study of seismic active arth pressure acted on gravity wall in backfill sand,"China Journal of Highway and Transport, vol. 22, no. 6, pp. 26–33, 2009.

17. R. Siddharthan, S. Ara, and G. Norris, "Simple rigid plastic model for seismic tilting of rigid walls,"Journal of structural engineering New York, N.Y., vol. 118, no. 2, pp. 469–487, 1992. ·

18. X. Zeng, "Seismic response of gravity quay walls. I: centrifuge modeling," Journal of Geotechnical and Geoenvironmental Engineering, vol. 124, no. 5, pp. 406–417, 1998.

19. M. Celebi, "Seismic responses of two adjacent buildings. II. Interaction," Journal of Structural Engineering, vol. 119, no. 8, pp. 2477–2492, 1993.

20. H. A. Smith and W.-H. Wu, "Effective optimal structural control of soil-structure interaction systems,"Earthquake Engineering & Structural Dynamics, vol. 26, no. 5, pp. 549–570, 1997.

21. N. M. Newmark, "Effects of earthquakes on dams and embankments," Geotechnical, vol. 15, no. 2, pp. 59–139, 1965.

22. Y. Z. Wang, "Distribution of earth pressure on a retaining wall," Geotechnique, vol. 50, no. 1, pp. 83–88, 2000.

23. R. Richards and D. G. Elms, "Seismic behavior of gravity retaining walls," Journal of Geotechnical Engineering, vol. 105, no. 4, pp. 444–449, 1979.

24. D. C. Yao, "Retaining wall seismic failure analysis," Engineering Journal of Wuhan University, vol. 36, no. 2, pp. 82–87, 2003.

25. J. H. Wood, "Earthquake-induced pressures on a rigid wall structure," Bulletin of the New Zealand Society for Earthquake Engineering, vol. 8, no. 3, pp. 86–175, 1975.

26. X. Zeng and R. S. Steedman, "Rotating block method for seismic displacement of gravity walls," Journal of Geotechnical and Geoenvironmental Engineering, vol. 126, no. 8, pp. 709–717, 2000. ·

27. B. Munwar Basha and G. L. Sivakumar Babu, "Reliability assessment of internal stability of reinforced soil structures: a pseudo-dynamic approach," Soil Dynamics and Earthquake Engineering, vol. 30, no. 5, pp. 336–353, 2010.

7
CHAPTER

ENHANCING SEISMIC CAPACITY OF PILE-SUPPORTED WHARVES USING YIELDING DAMPERS

Seyed Amin Mousavi and Khosrow Bargi

INTRODUCTION

During an earthquake event liquefaction of saturated loose sandy soils and excessive piles drifts make the most common causes of damages to pile-supported wharves. Therefore, in absence of liquefaction conditions, pile drift can be considered as a suitable indicator in order to evaluate performance of wharves under seismic events. Some techniques, which rely on stiffness increasing, such as inclined piles, have been investigated in earlier studies by Gerolymos et al. [1] and Poulos [2], as a method to reduce lateral displacements of pile-supported wharves. Inclined piles have two main drawbacks, high construction costs and punching failures in their connections. As reported by Oyenuga et al. [3], however, the punching failure problem can be moderated using a new design approach for pile-deck connections. Lehman et al. [4] have also improved performance of pile-wharf connections. In another study a novel stone column has been proposed by Mageau and Chin [5] to improve seismic behavior of wharves. Using passive control techniques, this study tried to improve the seismi behavior of pile-supported wharves.

Nowadays passive control methods have gained more attention in order to mitigate natural or man-made structural vibrations. Some of these passive techniques have been briefly described by Soong and Dargush [6]. Earlier studies on passive control techniques have been commonly restricted to long period structures, such as tall buildings, long-span bridges, and offshore jacket platforms, to the authors' knowledge, passive energy dissipation devices have not been investigated earlier for wharves. Among various passive dampers, metallic yielding damper seems to be more appropriate as yielding dampers are easy to manufacture and need no specific maintenance. To date, many studies have been devoted to yielding dampers [7–10]. However, they have mainly

Citation: Seyed Amin Mousavi and Khosrow Bargi, "Enhancing Seismic Capacity of Pile-Supported Wharves Using Yielding Dampers," Journal of Structures, vol. 2013, Article ID 329130, 13 pages, 2013. doi:10.1155/2013/329130.

focused on building structures in which dampers were in conjunction with chevron braces.

Nowadays, with increasing ship sizes, berth deepening seems to be inevitable and consequently a new generation of pile-supported wharves would have relatively larger periods. Accordingly, the focus of this study is mainly on flexible wharves. As depicted in Figure 1, stability conditions of the retaining wall would dictate the maximum allowable yielding force of the damper. Note that the yielding force of the damper can be adjusted to very small values. In other words, during a seismic event, damper forces can be tuned to be far less than those imposed by other sources, such as backfill soil pressure or inertial force of the retaining wall. As a result, the damper can be designed as if it has virtually no destructive effect on the retaining wall. It should be elaborated that the retaining wall can be of any type, for example, gravity wall, caisson, sheet pile, and so forth. During this study, metallic yielding damper, metallic damper, yielding damper, and damper are all considered to have the same meaning and are used instead of each other.

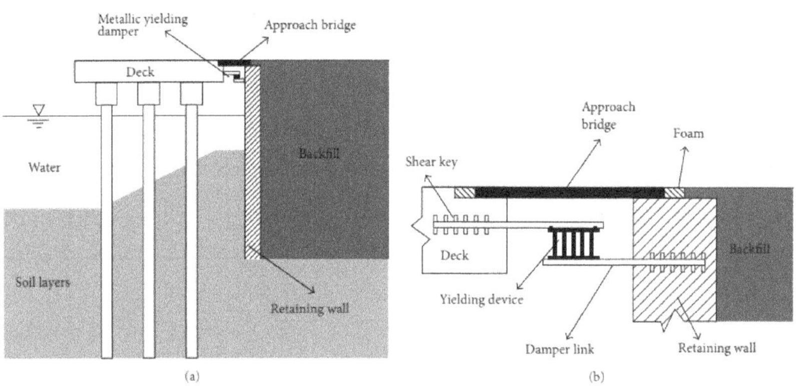

Figure 1: Damper-wharf system.

SCOPE OF THE CURRENT STUDY

The main objective of the present study is to mitigate seismic-induced responses of pile-supported wharves using metallic yielding dampers that were placed between the deck and the retaining wall (quay wall) in specific intervals along the shoreline, as illustrated in Figure 1. In the design procedure, it is very important to transmit the damper force to the ground within a safe load path. As a result, depending on residual capacity of the retaining wall, there is a restriction on the allowable yielding force of the damper. As shown in Figure 1, the damper includes two parts, the yielding device and the damper link. The damper link connects the yielding device to the wharf deck and the retaining wall. The link is very stiff due to its short lengths and can be assumed to be rigid. To optimize different parameters of the damper, sets of values for damper stiffness and yielding force are considered, and the contribution of the damper for each of them is evaluated.

OPTIMIZATION PROCEDURE

In the optimization process, four single degree of freedom (SDOF) systems with periods of 0.4 s, 0.6 s, 0.8 s, and 1.0 s are adopted and subjected to 20 digitized ground acceleration records. It is assumed that, during an earthquake, displacements at the top of the retaining wall are the same as those at the base of the wharf, that is, sea bed. In other words, displacements at the free end of the damper element are the same as those at the base of the SDOF system, Figure 2. Accordingly, during the optimization process, the retaining wall is not modeled as there are numerous different walls with different dynamic characteristics, and it is not practical to consider all possible wall conditions. Besides, this simplification is reasonable as the period of the wharf is far above than that of wall-soil system, especially in the case of long period wharves which are the main concern of this study. As a result, the wall-soil system can be assumed to be rigid compared to the wharf. Besides, the sea bed is very close to the top of the retaining wall, and it is possible to assume that seismic waves hit these two points simultaneously. While there are some experimental evidence that validate the previous assumptions [11, 12], in the Numerical Assessment, validity of this assumption would be also investigated. Three values of 0.05, 0.1, and 0.2 are adopted for yielding force-deck weight ratio, that is, the ratio of the damper yielding force to weight of the SDOF system. In addition, for each SDOF system, different stiffness ratios, that is, ratio of the damper stiffness to the SDOF stiffness, are adopted according to the period of the SDOF system, as presented in Table 1. Stiffness ratios for each period are selected so that maximum ductility ratio of the damper falls in a reasonable range.

Table 1: Stiffness ratios which are considered for the SDOF systems

Period of the SDOF system (s)	Considered stiffness ratios
0.4	0.5, 1, 2, 3, 4
0.6	1, 2, 3, 4, 5
0.8	2, 4, 6, 8, 10
1	4, 6, 8, 10, 12

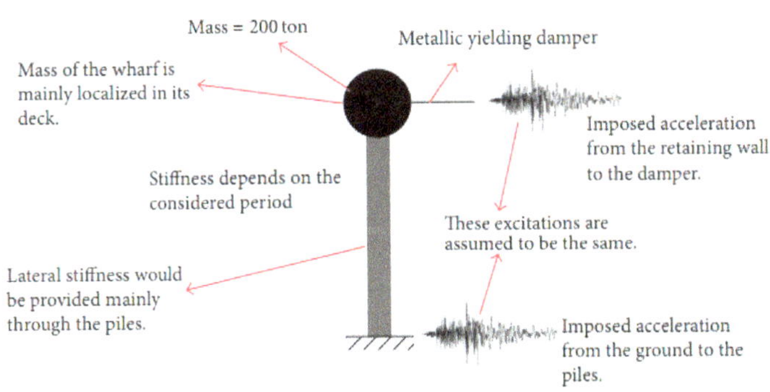

Figure 2: SDOF-damper system that is used in the optimization process. Mass = 200 ton.

Note that a constant value of 0.03 is considered for the strain hardening ratio of the damper material. As reported by Xia and Hanson [7], strain hardening ratio has a little influence on the behavior of yielding dampers. Therefore, the strain hardening is excluded from the carried out parametric study. According to most codes of practice, such as OCDI [13], pile-supported wharves should be designed to remain mainly elastic during ultimate seismic excitations. As a result, in the optimization procedure behaviors of the adopted SDOFs are assumed to be linear elastic.

RECORD SELECTION

As mentioned earlier, 20 digitized ground motion records, obtained from PEER database [14], are used in the optimization process. Adopted ground motions have magnitudes greater than 6, and they are selected such that their averaged Fourier spectrum covers a broad band of periods of interest, that is, between 0.4 s and 1 s. Characteristics of these ground motions are illustrated in the appendix. Moreover, frequency contents and acceleration response spectra of these 20 digitized ground acceleration records are shown in Figure 3. In the optimization process, all of these ground accelerations are scaled to 0.3 g.

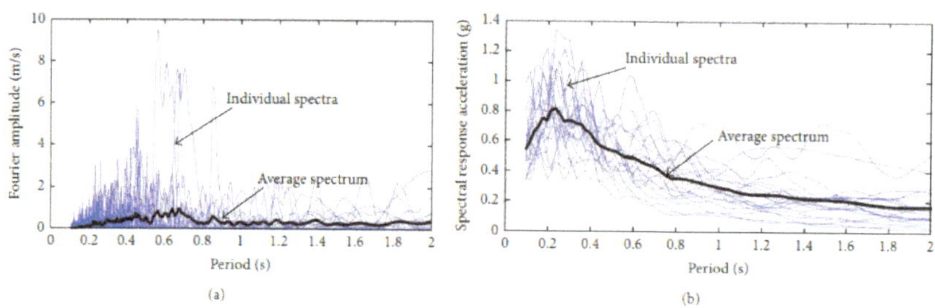

Figure 3: (a) Fourier spectra of the considered earthquakes and (b) 5% damped spectral response acceleration of the considered earthquakes.

It should be elaborated that considering the lower PGAs is not necessary as the purpose of the damper is to remain elastic during moderate ground motions, and the main concern is only about extreme earthquake events. Note that selected ground accelerations are not explicitly scaled according to their response spectrum the as effect of their frequency contents on the damper parameters would be also investigated during the parametric study. According to Figure 3 and Table 1, however, it is clear that most of the considered earthquakes would impose a high level of resonance on the considered SDOF systems which is in agreement with the main philosophy of the spectrum-based scaling technique.

OPTIMUM PARAMETERS OF THE YIELDING DAMPER

Optimization of the damper parameters is not a straightforward procedure. While higher damper stiffness would decrease deck displacement, it would also increase ductility demand of the damper. As a result, for a particular yielding force, higher stiffness values are not necessarily preferable due to the limitation on maximum allowable ductility of the damper.

In this study mean maximum displacement ratio of each SDOF system is considered as the optimization index, and the mean maximum ductility ratio is also obtained for different values of the damper parameters. The results are shown in Figures 4, 5, 6, and 7. Required analysis is carried out by OpenSees, which has been developed by Mazzoni et al. [15]. Giuffre-Menegotto-Pinto model [15] is also adopted to simulate nonlinear behavior of the yielding damper. It should be clarified that, for each SDOF system, mean maximum displacement ratio is defined to be the mean of the maximum displacement ratios with and without damper, and mean maximum ductility ratio is the mean of the maximum damper ductility ratios for all 20 ground motion records. Note that the ductility ratio is the ratio of the damper deformation to its yielding deformation such that it cannot be smaller than one.

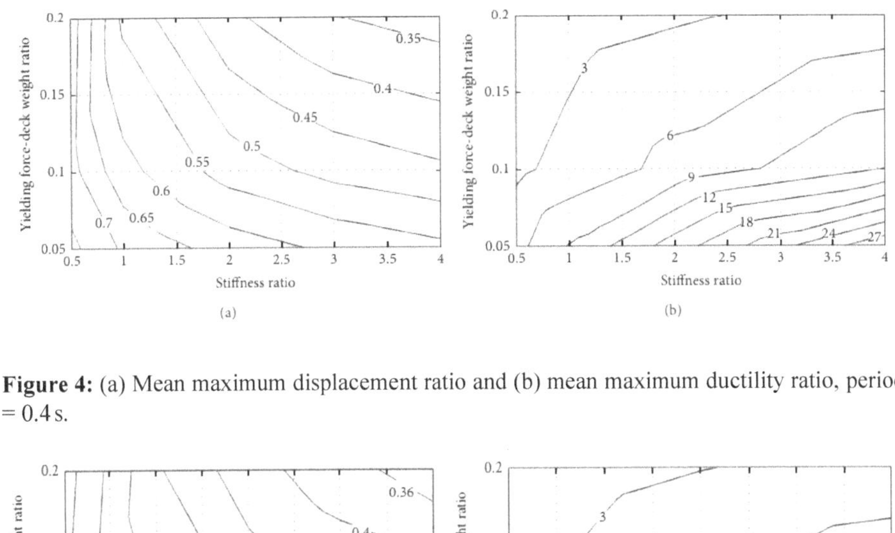

Figure 4: (a) Mean maximum displacement ratio and (b) mean maximum ductility ratio, period = 0.4 s.

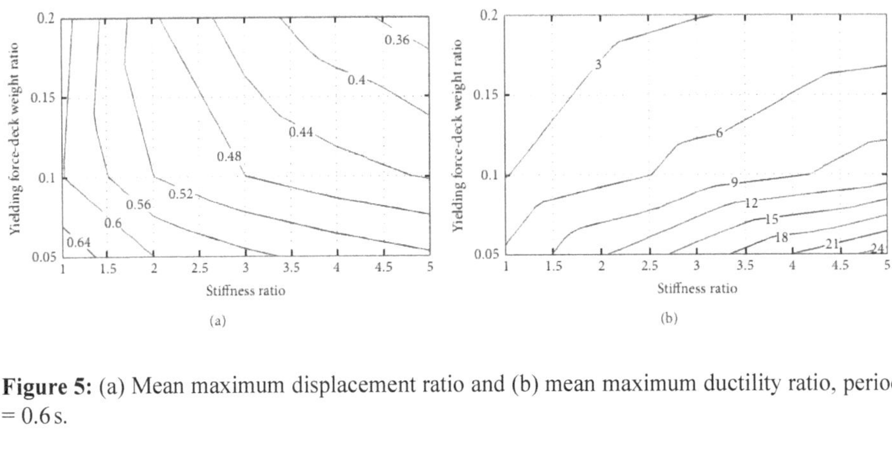

Figure 5: (a) Mean maximum displacement ratio and (b) mean maximum ductility ratio, period = 0.6 s.

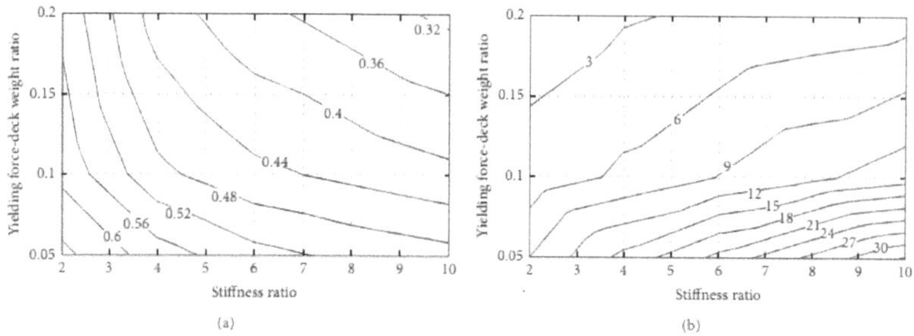

Figure 6: (a) Mean maximum displacement ratio and (b) mean maximum ductility ratio, period = 0.8 s.

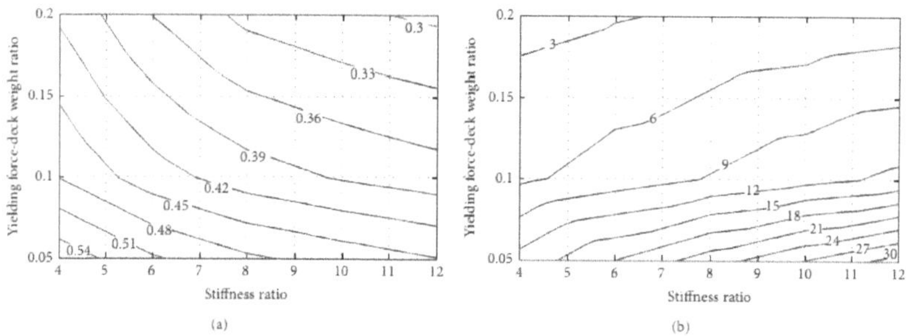

Figure 7: (a) Mean maximum displacement ratio and (b) mean maximum ductility ratio, period = 1.0 s.

As stated earlier, the retaining wall would determine the allowable yielding force of the damper. Therefore, optimum values of the damper parameters depend on the period of the soil-wharf system, target maximum ductility ratio and stability conditions of the retaining wall. According to Figures 4–7, in the case of short period wharves and for low stiffness ratios, by increasing the damper yielding force no improvement would occur on the damper efficiency. In general, the efficiency of the damper always increases by increasing the stiffness ratio; however, the maximum ductility ratio of the damper would also increase. Based on earlier experimental tests which have been reported by Xia and Hanson [7], value of about 6 is well suited for the maximum ductility ratio of a metallic yielding damper. As a result, other parameters should be selected such that maximum ductility ratio of the damper remains close to 6.

Using resonance index concept, the effects of the frequency content on the damper efficiency and also on the maximum ductility ratio are investigated. Resonance index, as defined by Mousavi et al. [16], is the ratio of the Fourier amplitude corresponding to the period of the system to the maximum Fourier amplitude. Accordingly, if the value of the resonance index is close to one, the structure would experience a high level of resonance, and, if its value is close to zero, no noticeable resonance would occur. It should be pointed out that the term resonance in the current study refers to resonance of the first mode, and it would happen if the frequency content of the ground acceleration has high Fourier amplitudes in frequencies close to the fundamental frequency of the structure.

As illustrated in Figure 8, the resonance index has affected the damper efficiency and its maximum ductility ratio in an irregular manner. This is mainly due to nonlinear behavior of the damper and different pulse arrangements of the considered records. In other words, during an earthquake event, a wharf-damper system does not have a constant period, and its period would be a time, dependent parameter. From Figure 8 it is obvious that, in a general manner, by increasing the resonance index, the maximum ductility ratio increases, and the RMS (root mean square) displacement ratio decreases. This result is quite reasonable as in the resonance condition the damper would have a greater effect on the response and consequently would suffer greater damages (larger ductility demands).

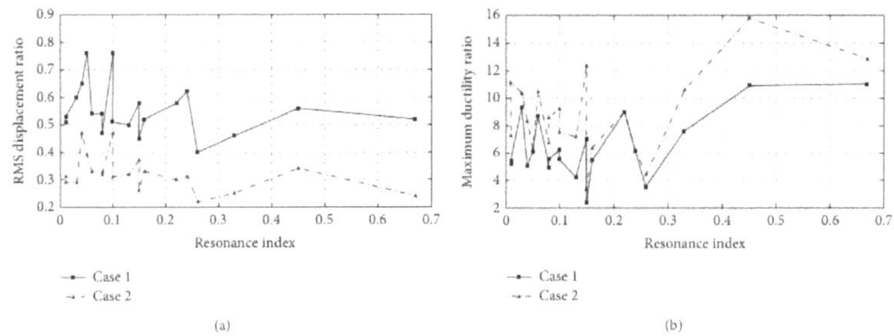

Figure 8: Dependency of (a) damper efficiency and (b) damper maximum ductility ratio on frequency content of the ground motions. Case 1: period = 0.6 s, stiffness ratio = 1, and yielding force-deck weight ratio = 0.05. Case 2: period = 0.6 s, stiffness ratio = 4, and yielding force-deck weight ratio = 0.10.

According to the obtained results, using Matlab's Curve Fitting Toolbox [17], following relations for the stiffness ratio and its corresponding mean maximum displacement ratio can be proposed. Note that in proposed relations the allowable maximum ductility ratio is restricted to 6:

$$\mathrm{SR} = 0.875e^{14.2\alpha}T, \tag{1}$$

$$\mathrm{MR} = (0.9 - 3.15\alpha)(1 - 0.15T), \tag{2}$$

where SR is the stiffness ratio, MR represents the maximum displacement ratio, α denotes the yielding force-deck weight ratio, and T stands for the period of the soil-wharf system (without damper).

In the case of yielding dampers, structural response would be reduced due to two reasons: (a) the energy dissipation capability of the damper and (b) stiffness increasing due to stiffness of the damper. While, in conventional methods [6, 18], both of the previous effects have been considered in estimating the equivalent viscous damping, in this study a new approach is proposed for evaluating the equivalent viscous damping. In this method, using the maximum displacement ratio, MR, and relevant response spectra, such as those proposed in FEMA-356 [19], the equivalent viscous damping can be estimated. According to the definition of MR, it is possible to rewrite it as follows:

$$\text{MR} = \frac{S_{a_{\text{eq}}}}{S_a}, \tag{3}$$

where Sa_{eq} and S_a are the spectral response accelerations of the wharf with and without damper, respectively. Having MR and S_a, it is possible to obtain value of the Sa_{eq} and consequently the equivalent viscous damping. Obviously, the equivalent viscous damping can be used only for preliminary design, and for the final design nonlinear behavior of the damper should be explicitly considered.

DESIGN OF THE DAMPER

A simple retrofit-based design procedure is proposed in this section assuming that the stability condition of the retaining wall cannot be improved. A stable condition for a sheet pile wall would be provided if its safety factors in shear and flexure remain in acceptable ranges. In the case of a gravity wall, stability condition would be provided if the wall has acceptable safety factors against slip and rotation. The authors would like to elaborate that the damper force has no noticeable effect on other failure modes, such as foundation bearing or settlement.

Step 1. Determine the allowable force that the retaining wall can sustain at the damper location during the design earthquake, R.

Again it should be clarified that R is a small fraction of redundant capacity of the retaining wall that is provided by its initial safety factor. It is obvious that reduction of the wall safety factor is inevitable, but the value of R can be tuned such that the reduced safety factor still remains in an acceptable range. Clearly, the damper force and other imposed forces on the wall are not necessarily in phase. However, in this design guideline, these forces are assumed to be in phase in favor of safety. A simplified guideline is presented in Section 6.1 for selecting the value of R.

Step 2. Determine the yielding force of the damper:

$$F_y = \text{FS} \cdot R, \tag{4}$$

where F_y represents yielding force of the damper and FS is an appropriate factor of safety lower than the unity which considers the effect of the damper strain hardening (value of 0.9 is reasonable for FS.)

Step 3. Determine the value of the yielding force-deck weight ratio, α:

$$\alpha = \frac{F_y}{W_D}, \tag{5}$$

where W_D is the seismic weight of the deck attributed to the damper.

Step 4. Determine the period of the soil-wharf system, T.

Step 5. According to (1), evaluate the stiffness ratio, SR, and consequently damper stiffness as follows:

$$K = \frac{4\pi^2 W_D \mathrm{SR}}{gT^2}, \tag{6}$$

where g and K stand for acceleration of gravity and damper stiffness, respectively.

Step 6. For the preliminary design, according to (2) and (3), estimate the obtained equivalent viscous damping, and for the final design use the obtained yielding force (Step 2) and stiffness (Step 5) of the damper in a nonlinear model.

There are some relations in the literature relating stiffness of the yielding damper to its geometrical parameters, such as those proposed by Tehranizadeh [20] for added damping and stiffness devices (ADAS) and those obtained by Soong and Dargush [6] for triangular added damping and stiffness dampers (TADAS). As mentioned earlier, stiffness and yielding displacements of the yielding device are equal to those of the yielding damper due to the rigidity of the damper link.

It should be pointed out that, if the stability condition of the retaining wall can be improved, the design direction would be changed. In this case, according to the desired damping ratio, value of MR can be evaluated first, and then required values of α and SR would be obtained according to (2) and (1), respectively.

R Selection.

A simple yet accurate guideline is presented here to estimate the appropriate value for R. Note that all of the presented calculations should be carried out for the most critical section of the wall which experiences the largest forces and has the lowest capacity.

Step 1. Determine shear and flexural (or slip and rotational) capacities of the retaining wall, C_V and C_f, respectively, at the critical section of the wall.

Step 2. Estimate the maximum expected seismic inertial force of the wall, F_I, the maximum seismic horizontal force from the backfill soil, F_s, the hydrostatic force, F_{hs}, and the maximum hydrodynamic force, F_{hd}. It should be pointed out that F_s and F_{hd} can be evaluated by many techniques such as Mononobe-Okabe method and Westergaard's theory, respectively [21]. Note that F_{hs} tends to stabilize the wall while F_{hd} could have a destructive effect depending on its direction.

Step 3. Estimate the maximum moments at the critical section of the wall due to its inertial force, M_I, the backfill soil pressure, M_s, the hydrostatic force, M_{hs}, and the hydrodynamic force, M_{hd}.

Step 4. Determine the acceptable reduced factor of safety, FS_r, of the retaining wall which is greater than unity. As stated earlier, reduction of the retaining wall safety factor cannot be avoided. It is clear that this reduction must be in an allowable and reasonable range. In fact just a 10% reduction is enough for a significant reduction in the wharf response.

Step 5. Select the minimum of R_V and R_M as the value of R. Equation (7) is obtained based on the shear and moment equilibrium of the retaining wall in which R_V and R_M are the maximum allowable damper forces according to the shear and flexural (or slip and rotationl) capacity of the retaining wall, respectively:

$$R_V = \frac{C_V}{\mathrm{FS}_r} - F_I - F_S - F_{\mathrm{hd}} + F_{\mathrm{hs}},$$

$$R_M = \frac{1}{h}\left(\frac{C_f}{\mathrm{FS}_r} - M_I - M_S - M_{\mathrm{hd}} + M_{\mathrm{hs}}\right), \tag{7}$$

where h is the distance between damper-wall connection and the critical section of the wall. As the height of the wall increases, amounts of its external forces would also increase. Accordingly, the sensitivity of the wall on the damper force would decrease in the cases of higher quay walls.

NUMERICAL ASSESSMENT

Contribution of the proposed damper is evaluated in this section for a typical pile-supported wharf. Effect of the yielding damper on the retaining wall stability is also examined. Material and geometrical nonlinearities are not considered as structural elements are intended to remain mainly elastic per OCDI [13]. It should be noted that, in this example, soil-pile interaction is considered to examine validity of the optimization assumptions. Soil characteristics are considered based on a real soil profile; however, they are not presented here due to space limitation. As depicted in Figure 9, three steel piles with an outer diameter of 1 m and wall thickness of 10 mm are assumed to support the deck. The piles are spaced per 5 m along the shoreline. Moreover, the adopted deck is a concrete slab with thickness of 0.4 m and seismic weight of 1600 kN. Slab length is 12 m perpendicular to the shoreline, and its width along the shoreline is assumed to be 5 m. In other words, a 5 m part of the wharf is separated and considered during this example, and value of 5% is considered as its inherent first mode damping ratio. It is assumed that the stability condition of the retaining wall in 5 m intervals is such that 10% of the deck weight can be considered as the damper yielding force. The effect of this value on the safety factor of the wall will be investigated during this example.

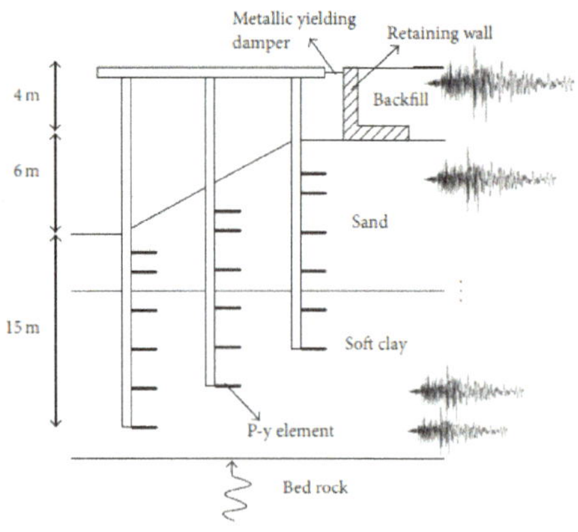

Figure 9: Considered soil-wharf-damper system.

Well-known p-y elements, as suggested by Boulanger et al. [22], are used to simulate soil-pile interaction, and pile group effect is also considered as proposed by Castelli and Maugeri [23]. Moreover, EERA, developed by Bardet et al. [24], is used for the required free field response analysis. It should be clarified that p-y element technique is a simplified method which cannot directly account for topographic irregularities. Besides, as suggested by OCDI [13], required inputs of this method are the obtained results of the free field analysis which is an equivalent-linear method, and its application is limited to soil strains of 1% or less.

The fundamental period of the fixed base wharf is 0.43 s, and, by considering the soil-pile interaction, its value would increase to 0.84 s. It is clear that 0.84 s should be used to estimate optimum parameters of the damper. As a result, in this assessment, α and T are 0.1 and 0.84, respectively, and SR and MR, respectively, would be 3 and 0.51. Northridge 1994, Parkfield 1966, and Kobe 1995 ground motions are used in this example. The so-called earthquakes are scaled to PGA = 0.3 g at rock outcropping. As shown in Figure 10, these seismic events are adopted due to their difference in the frequency contents at the period of 0.84 s. Note that adopted ground motions are not scaled based on their response spectrum as this assessment intends to examine the resonance effect on the damper contribution as well.

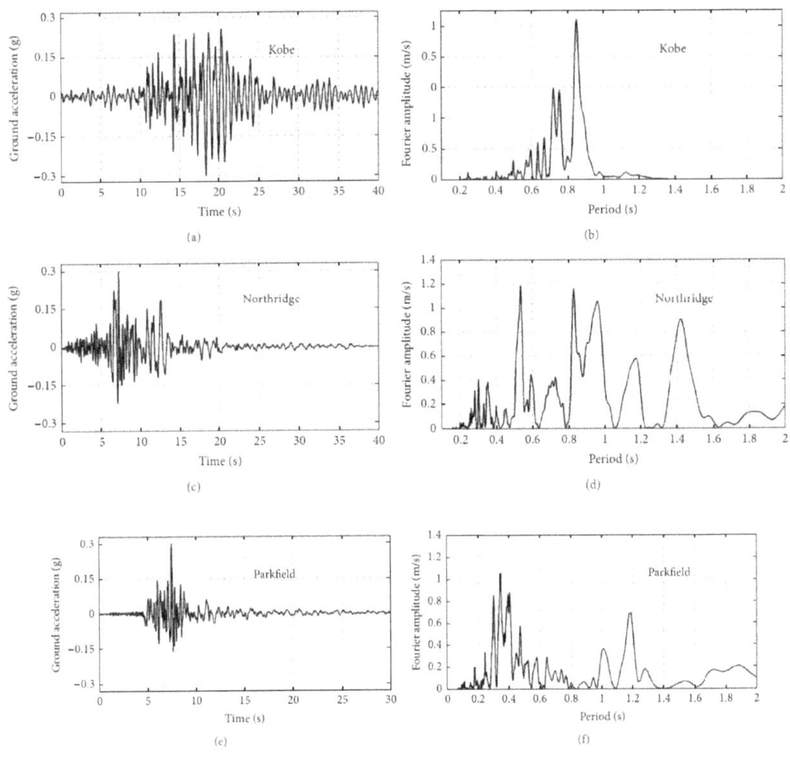

Figure 10: Considered ground accelerations at the rock outcropping and their corresponding Fourier spectra.

Obviously, for a practical design, these ground motions should be scaled based on an applicable code.

As suggested by Takahashi and Takemura [25], one of the most important reported wharf damages was bending moment on piles due to their large lateral drifts. Therefore, in this example, drift of the middle pile is considered as the response of the structure and is compared in different cases. Figures 11–13 and Table 2represent the obtained results. It should be elaborated that the optimum stiffness of the damper is estimated according to (1). It is clear that the damper effectively reduced the response of the wharf and saved the piles from severe seismic-induced damages, regardless of the resonance condition.

Table 2: Results of the numerical example

Earthquake event	Without damper		With damper		
	Maximum middle pile drift (cm)	RMS middle pile displacement (cm)	Maximum middle pile displacement (cm)	RMS middle pile displacement (cm)	Maximum damper ductility ratio
Northridge	6.62	2.76	3.35	0.99	5.7
Parkfield	1.59	0.60	1.47	0.35	2.5
Kobe	9.51	4.99	5.94	2.51	10.1

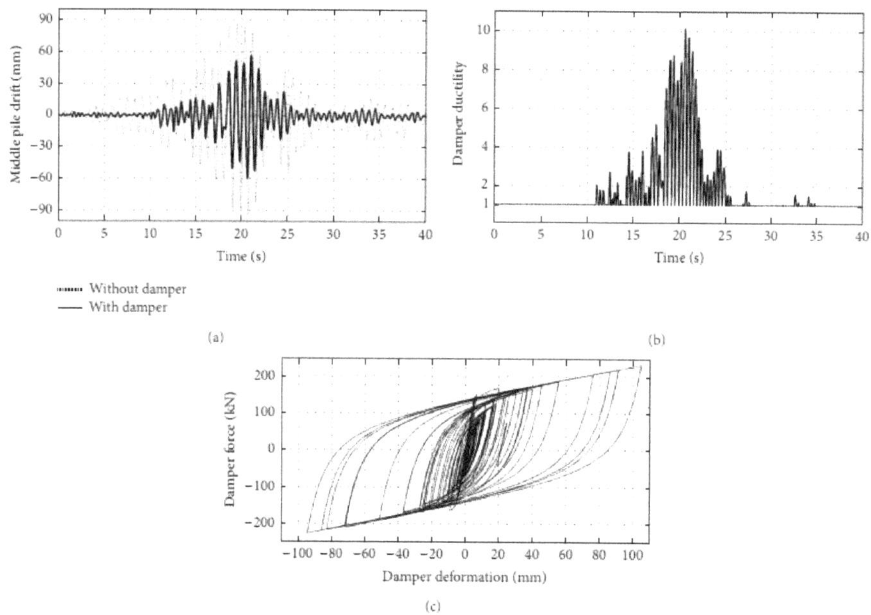

Figure 11: (a) Drift of the middle pile with and without damper, (b) ductility ratio of the damper, and (c) cyclic behavior of the damper-Kobe 1995 earthquake.

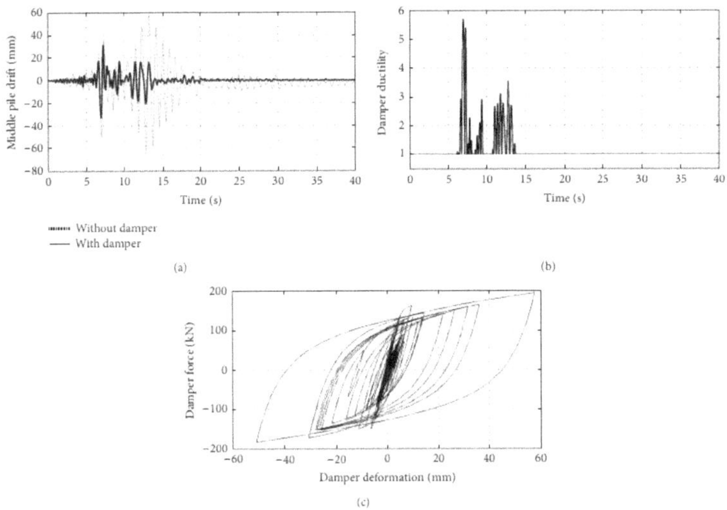

Figure 12: (a) Drift of the middle pile with and without damper, (b) ductility ratio of the damper, and (c) cyclic behavior of the damper-Northridge 1994 earthquake.

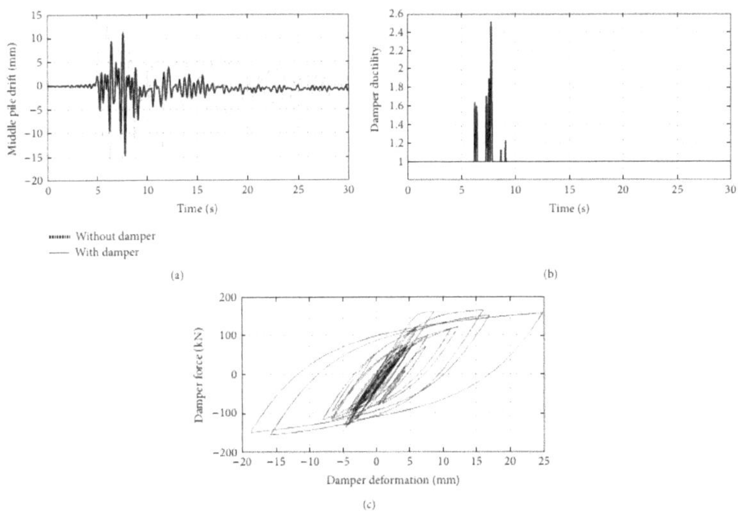

Figure 13: (a) Drift of the middle pile with and without damper, (b) ductility ratio of the damper, and (c) cyclic behavior of the damper-Parkfield 1966 earthquake.

It can be seen from Table 2 that, in the resonance condition, that is, Northridge and Kobe events, the damper effectively reduced both maximum and RMS responses, and the reduction was mainly due to the energy dissipation capability of the damper. However, in the case in which the resonance has not occurred, that is, Parkfield earthquake, the damper was able to reduce only the RMS response mainly through stiffness increasing rather than energy dissipation. This should be considered as the merit of yielding dampers as these devices remain mainly elastic during service level ground motions and localize damages and show noticeable energy dissipation capabilities in the cases of design level earthquakes. In Kobe earthquake, ground acceleration frequencies are concentrated in a narrow band making the earthquake very similar to a harmonic excitation with the resonance frequency. Therefore, the damper sustained large ductility demands during the Kobe ground motion. However this value is still in a safe range. Table 3 represents validity of the optimization assumptions. Parkfield earthquake was a very weak excitation for the considered wharf and is excluded in Table 3. Note that intensities of the earthquakes are evaluated based on their spectral acceleration spectra as shown in Figure 14.

Table 3: Validity of the optimization assumptions

Earthquake event	Earthquake intensity	Maximum displacement ratio (MR)		Maximum ductility ratio	
		Predicted	Observed	Predicted	Observed
Northridge	strong	0.51	0.51	6	5.7
Kobe	Very strong	0.51	0.62	6	10.1
Parkfield	weak	0.51	0.92	6	2.5

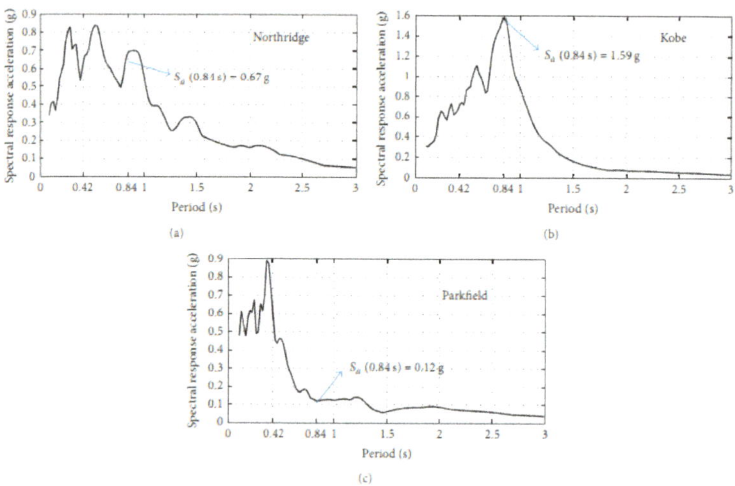

Figure 14: Spectral response accelerations of the considered ground motions.

Obtained results indicate that the proposed technique is accurate for design level earthquakes (Northridge earthquake in this case). However the accuracy would be lost for very weak and very strong ground motions.

Friction angle, density, and pore pressure ratio of the backfill soil are assumed to be 30 degree, 1800 kg/m^3, and 0.15, respectively. Dead and live loads on the backfill soil are 35 kN/m^2. Moreover, thickness of the retaining wall is 0.8 m. As a result, the maximum slip force and rotational moment on the wall would be 580 kN and 1110 kN·m per unit length of the wall, respectively. As illustrated earlier in Figure 12, the maximum damper force in the case of design level earthquake (Northridge earthquake) is only

200 kN for 5 m length of the wall, that is, 40 kN per unit length, and its corresponding rotational moment is 140 kN·m per unit length. It is obvious that the maximum damper force is much smaller than that imposed from other sources.

If initial safety factor of the wall was assumed to be 1.5, the reduced safety factor would be about 1.33. Therefore by 11% reduction in the initial safety factor of the retaining wall, maximum response of the wharf would be decreased by 49%. In other words, the proposed damper would shift the safety factor from the retaining wall to the piles which are more vulnerable elements. Again it should be clarified that different forces are not necessarily in-phase, and the real reduced safety factor of the wall would be larger than 1.33 and closer to its initial value.

CONCLUSIONS

Through this study contribution of yielding damper on reducing seismic-induced responses of pile-supported wharves was examined. Different parameters of the proposed yielding damper were also optimized. Obtained results indicate that yielding damper can dissipate a considerable portion of the input energy and effectively suppress pile drifts and save them from serious damages during seismic events. Yielding damper has a minor destructive effect on the retaining wall, and selecting an appropriate value for its yielding force is a crucial task. In general, it was shown that the optimum parameters of the damper depend on the retaining wall stability condition, period of the soil-wharf system, and allowable maximum ductility ratio of the damper.

As stated earlier, the damper would inevitably reduce the safety factor of the retaining wall, but this reduction is very small and can be neglected. For example, in the presented numerical example, this reduction was about 11%, and the reduction in response of the wharf was about 50%. In other words, yielding dampers slightly reduce the safety factor of the retaining wall, whereas effectively increase safety factor of the piles which are much more vulnerable to seismic excitations. It should be pointed out that the so-called destructive effect would be smaller in the case of higher retaining walls. The numerical example also indicates that metallic yielding dampers can effectively suppress RMS response of pile-supported wharves during any kind of ground motion regardless of its intensity.

It should be noted that, according to the optimization assumptions and damper placement, flexible wharves with solid retaining walls are the best options for the presented technique. This technique is not recommended for wharves with weak retaining walls due to the used philosophy, that is, shifting the safety factor from the wall to the wharf piles.

The authors would like to elaborate that there are still some unanswered questions about the proposed technique that call for further studies. For example, if the wall-soil system behind the wharf was very soft, some of the adopted optimization assumptions may not be valid anymore. Besides, as suggested during the paper, imposed damper force on the retaining wall is out of phase with other forces, and this feature can be formulated in other studies. Consideration of this effect would decrease the destructive effect of the proposed technique on the retaining wall.

Appendix

Ground motion records that were used in the optimization process are presented in Table 4.

Table 4: Ground motion records used in the optimization process

Earthquake	Station name	Comp.	Mag.	PGA (*g*)	Source
Borrego mountain 04/09/68	El centro array #9	180	6.8	0.13	Usgs
Borrego mountain 04/09/68	El centro array #9	270	6.8	0.057	Usgs
Imperial valley 5/19/40	El centro array #9	180	7	0.313	Usgs
Imperial valley 5/19/40	El centro array #9	270	7	0.215	Usgs
Kern county 7/21/52	Taft lincoln school	021	7.4	0.156	Usgs
Kern county 7/21/52	Taft lincoln school	111	7.4	0.178	Usgs
San fernando 02/09/71	La hollywood stor lot	090	6.6	0.21	Usgs
San fernando 02/09/71	La hollywood stor lot	180	6.6	0.174	Usgs
San fernando 02/09/71	Lake hughes #4	111	6.6	0.192	Usgs
San fernando 02/09/71	Lake hughes #4	201	6.6	0.153	Usgs
Duzce 11/12/99	Lamont station 1062	N	7.1	0.114	Lamont
Duzce 11/12/99	Lamont station 1062	E	7.1	0.257	Lamont
Northridge 01/17/94	Canyon country-w lost canyon	000	6.7	0.41	Usc
Northridge 01/17/94	Canyon country-w lost canyon	0270	6.7	0.482	Usc
Northridge eq 1/17/94	Lakewood-del amo	000	6.7	0.137	Usc
Northridge eq 1/17/94	Lakewood-del amo	090	6.7	0.123	Usc
Parkfield 06/28/66	Cholame #5	085	6.1	0.442	Cdmg
Parkfield 06/28/66	Cholame #5	355	6.1	0.367	Cdmg
Manjil, Iran	Station code: 18	,-N66E	7.37	0.184	—
Manjil, Iran	Station code: 18	,-N24W	7.37	0.131	—

REFERENCES

1. N. Gerolymos, A. Giannakou, I. Anastasopoulos, and G. Gazetas, "Evidence of beneficial role of inclined piles: observations and summary of numerical analyses," Bulletin of Earthquake Engineering, vol. 6, no. 4, pp. 705–722, 2008.
2. H. G. Poulos, "Raked piles—virtues and drawbacks," Journal of Geotechnical and Geoenvironmental Engineering, vol. 132, no. 6, pp. 795–803, 2006.
3. D. Oyenuga, E. Abrahamson, A. Krimotat, A. Kozak, T. Labasco, and F. Lobedan, "A study of the pile-wharf deck connection at the Port of Oakland," in Ports 2001: American's Ports-Gateways to the Global Economy, 2004.
4. D. E. Lehman, E. Brackmann, A. Jellin, and C. W. Roeder, "Seismic performance of pile-wharf connections," in Proceedings of the ASCE Technical Council on Lifeline Earthquake Engineering Conference (TCLEE '09), p. 84, Oakland, Calif, USA, July 2009.
5. D. Mageau and K. Chin, "Effectiveness of stone columns on slope deformations beneath wharves," inProceedings of the ASCE Technical Council on Lifeline Earthquake Engineering Conference (TCLEE '09), p. 94, Oakland, Calif, USA, July 2009.
6. T. T. Soong and G. Dargush, Passive Energy Dissipation Systems in Structural Engineering, Wiley, Chichester, UK, 1997.
7. C. Xia and R. D. Hanson, "Influence of ADAS element parameters on building seismic response," Journal of Structural Engineering, vol. 118, no. 7, pp. 1903–1918, 1992.

8. D. Foti, L. Bozzo, and F. Lopez-Almansa, "Numerical efficiency assessment of energy dissipaters for seismic protection of buildings," Earthquake Engineering and Structural Dynamics, vol. 27, pp. 543–556, 1998.
9. L. M. Moreschi and M. P. Singh, "Design of yielding metallic and friction dampers for optimal seismic performance," Earthquake Engineering and Structural Dynamics, vol. 32, no. 8, pp. 1291–1311, 2003.
10. F. C. Ponzo, A. di Cesare, D. Nigro et al., "JET-PACS project: dynamic experimental tests and numerical results obtained for a steel frame equipped with hysteretic damped chevron braces," Journal of Earthquake Engineering, vol. 16, pp. 662–685, 2012.
11. I. Towhata, M. J. Alam, T. Honda, and S. Tamate, "Model tests on behaviour of gravity-type quay walls subjected to strong shaking," Bulletin of the New Zealand Society for Earthquake Engineering, vol. 42, no. 1, pp. 47–56, 2009.
12. A. Sadrekarimi, A. Ghalandarzadeh, and J. Sadrekarimi, "Static and dynamic behavior of hunchbacked gravity quay walls," Soil Dynamics and Earthquake Engineering, vol. 28, no. 2, pp. 99–117, 2008.
13. OCDI, Technical Standards and Commentaries for Port and Harbor Facilities in Japan, The Overseas Coastal Area Development Institute of Japan, Tokyo, Japan, 2002.
14. PEER, Strong Ground Motion Database, 2009, http://peer.berkeley.edu/smcat/.
15. S. Mazzoni, F. McKenna, M. H. Scott, and G. L. Fenves, The OpenSees Command Language Manual: Version 1.7.3, Pacific Earthquake Engineering Center, University of California, Berkeley, Calif, USA, 2006.
16. S. A. Mousavi, K. Bargi, and S. M. Zahrai, "Optimum parameters of tuned liquid column-gas damper for mitigation of seismic-induced vibrations of offshore jacket platforms," Structural Control and Health Monitoring, vol. 20, no. 3, pp. 422–444, 2013.
17. Matlab, User Guide, Curve Fitting Toolbox, MathWorks Inc., Version 7.6.0, 2008.
18. Applied Technology Council (ATC), "Seismic evaluation and retrofit of concrete buildings," Tech. Rep. ATC 40, Applied Technology Council, Redwood City, Calif, USA, 1996.
19. ASCE, Prestandard and Commentary for the Seismic Rehabilitation of Buildings (FEMA356), Federal Emergency Management Agency, Washington, DC, USA, 2000.
20. M. Tehranizadeh, "Passive energy dissipation device for typical steel frame building in Iran," Engineering Structures, vol. 23, no. 6, pp. 643–655, 2001.
21. S. L. Kramer, Geotechnical Earthquake Engineering, Prentice Hall, Englewood Cliffs, NJ, USA, 1996.
22. R. W. Boulanger, C. J. Curras, B. L. Kutter, D. W. Wilson, and A. Abghari, "Seismic soil-pile-structure interaction experiments and analyses," Journal of Geotechnical and Geoenvironmental Engineering, vol. 125, no. 9, pp. 750–759, 1999.

23. F. Castelli and M. Maugeri, "Simplified approach for the seismic response of a pile foundation," Journal of Geotechnical and Geoenvironmental Engineering, vol. 135, no. 10, pp. 1440–1451, 2009.
24. J. P. Bardet, K. Ichii, and C. H. Lin, A Computer Program for Equivalent-Linear Earthquake Site Response Analyses, Department of Civil Engineering, University of Southern California, 2000.
25. A. Takahashi and J. Takemura, "Liquefaction-induced large displacement of pile-supported wharf," Soil Dynamics and Earthquake Engineering, vol. 25, no. 11, pp. 811–825, 2005.

8

CHAPTER

OPTIMUM DESIGN OF GRAVITY RETAINING WALLS USING CHARGED SYSTEM SEARCH ALGORITHM

S. Talatahari, R. Sheikholeslami, M. Shadfaran and M. Pourbaba

INTRODUCTION

Every time a product is created or designed to satisfy human needs, the creator tries to achieve the best solution for the task in hand and therefore performs optimization. This process is often manual, time consuming and involves a step by step approach to identify the right combination of the product and associated process parameters for the best solution. Often the manual approach does not allow a thorough exploration of the solution space to find the optimum design, resulting in suboptimal designs [1–3]. Therefore experienced engineers may be able to come up with solutions that fulfill some of the requirements on structural response, cost, aesthetics, and manufacturing but they will seldom be able to come up with the optimal structure.

One type of optimization methods is known as metaheuristic algorithms. These methods are suitable for global search due to their capability of exploring and finding promising regions in the search space at an affordable time. Meta-heuristic algorithms tend to perform well for most of the optimization problems [4–7]. As a new meta-heuristic approach, this paper utilizes charged system search algorithm (CSS) for the optimum design of gravity retaining walls subjected to seismic loading. Retaining walls is generally classified as gravity, semigravity (or conventional), nongravity cantilevered, and anchored. Gravity retaining walls are the walls which use their own weight to resist the lateral earth pressures. The main forces acting on gravity retaining walls are the vertical forces from the weight of the wall, the lateral earth pressure acting on the back face and the seismic loads. These forces are used herein to illustrate the design principles. If other forces are encountered, such as vehicular loads, they must also be included in the analysis. The lateral earth pressure is usually calculated by the Coulomb equation.

Citation: S. Talatahari, R. Sheikholeslami, M. Shadfaran, and M. Pourbaba, "Optimum Design of Gravity Retaining Walls Using Charged System Search Algorithm," Mathematical Problems in Engineering, vol. 2012, Article ID 301628, 10 pages, 2012. doi:10.1155/2012/301628.

The paper is structured as follows. After this introduction, Section 2 recalls the optimization problem statement. Then review of CSS is presented in Section 3. Test case is presented in Section 4 while optimization and sensitivity analysis results are reported and discussed. Finally, Section 5 summarizes the main findings of this study, and conclusion is drawn based on the reported results.

THE OPTIMIZATION PROBLEM

Gravity walls derive their capacity to resist lateral loads through dead weight of the wall. The earliest method for determining the combined static and dynamic earth pressure on a retaining wall was developed by Okabe [8] and Mononobe [9]. This method, generally referred to as the Mononobe-Okabe method, is based on plasticity theory and is essentially an extension of the Coulomb sliding wedge theory in which the transient earthquake forces are represented by an equivalent static force. Therefore the effect of the earthquake motion can be represented as inertial forces $K_h W$ and $K_v W$ acting at the centre of gravity of the mass [10]. The principle of this method is illustrated in Figure 1. The Mononobe-Okabe method was originally developed for a dry cohesion less material with the following two assumptions.(1)The wall yields sufficiently such that a triangular soil wedge behind the wall is formed at the point of incipient failure, with the maximum shear strength mobilized along the sliding surface.(2)The wall and the soil behave as a rigid body with the shear wave travelling at an infinite speed such that the acceleration effectively becomes uniform throughout the mass of the soil wedge. The expression of the total dynamic force, P_{AE} (Figure 1) is given below:

$$P_{AE} = \frac{1}{2}\gamma H^2(1 + K_v)C_{AE},$$

$$C_{AE} = \frac{\cos^2(\phi' - \theta - \beta)}{\cos\theta\cos^2\beta\cos(\delta + \beta + \theta)\left[1 + \sqrt{\sin(\phi' + \delta)\sin(\phi' - \theta - i)/\cos(\delta + \beta + \theta)\cos(i - \beta)}\right]^2}. \tag{2.1}$$

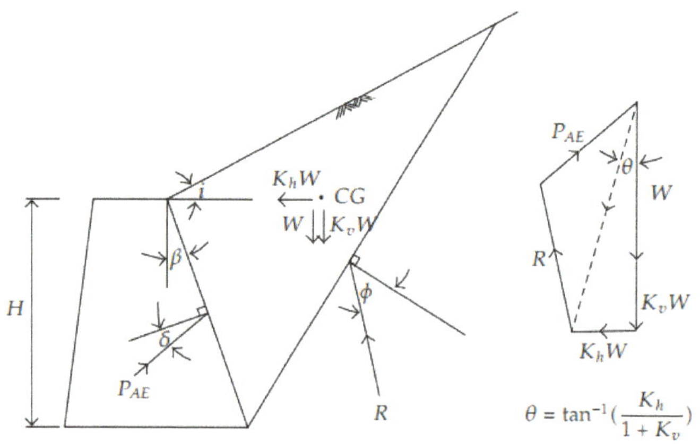

Figure 1: The Mononobe-Okabe method.

The pseudostatic approach can be visualized as effectively tilting the ground profile and wall geometry by an angle 0 defined as above, with a new gravity, g', given by the following equation:

$$g' = \sqrt{(1 + K_v)^2 + K_h{}^2}g. \tag{2.2}$$

It should be noted that the Mononobe-Okabe equation is applicable for retaining walls where the angle i is less than or equal to $(\phi' - \theta)$. This is because if the angle i is greater than $(\phi' - \theta)$, the sloping backfill behind the wall will be unstable unless the soil has sufficient cohesive strength. In the latter case, the more versatile analysis approaches should be adopted.

More advanced methods, such as dynamic response analysis and finite element method, are capable of allowing for the dynamic characteristics of the soil-structure system. However, these advanced methods are usually not justified for the analysis of conventional gravity retaining walls subjected to earthquake loading and the above simple methods are generally adequate as shown in [11]. Therefore, Mononobe-Okabe method is used herein to determine the dynamic earth pressure.

On the other hand, there are three different modes of instabilities, namely sliding, overturning, and bearing capacity, which should be checked [12]. The procedure for computing the dynamic factors of safety against sliding and overturning is same as that for static calculations, except that the inertia of the gravity wall itself must also be included when earthquake loading is considered [13]. Thus, the optimal seismic design problem of gravity retaining walls may be expressed as

Design variables

$$\mathbf{X} = [x_1, x_2, \ldots, x_6] \tag{2.3}$$

Minimize

$$W(\mathbf{X}) = A_{cs} \cdot \gamma \tag{2.4}$$

Constraints

$$FS_o \geq 2,$$

$$FS_s \geq 1.2,$$

$$FS_b \geq 3, \tag{2.5}$$

where X is the vector containing the design variables see Figure 2; W is the weight of a unit length of wall; A_{cs} is the wall cross-section area; γ is the density of the material; FS_o, FSs, and FS_b are the factors of safety against overturning, sliding, and bearing capacity, respectively.

CHARGED SYSTEM SEARCH ALGORITHM

The Charged System Search CSS algorithm is based on the Coulomb and Gauss laws from electrical physics and the governing laws of motion from the Newtonian mechanics. This algorithm can be considered as a multiagent approach, where each agent is a Charged Particle (CP). Each CP is considered as a charged sphere with radius a, having a uniform volume charge density and is equal to

$$q_i = \frac{\text{fit}(i) - \text{fitworst}}{\text{fitbest} - \text{fitworst}} \quad i = 1, 2, \ldots, N. \tag{3.1}$$

CPs can impose electric forces on the others, and its magnitude for the CP located in the inside of the sphere is proportional to the separation distance between the CPs, and for a CP located outside the sphere is inversely proportional to the square of the separation distance between the particles. The kind of the forces can be attractive or repelling, and it is determined by using ar_{ij}, the kind of force parameter, defined as

$$ar_{ij} = \begin{cases} +1 & \text{w.p. } k_t, \\ -1 & \text{w.p. } 1 - k_t, \end{cases} \tag{3.2}$$

where ar_{ij} determines the type of the force, 1 represents the attractive force, −1 denotes the repelling force, and k_t is a parameter to control the effect of the kind of force. In general, the attractive force collects the agents in a part of search space and the repelling force strives to disperse the agents. Therefore, the resultant force is redefined as

$$\mathbf{F}_j = q_j \sum_{i, i \neq j} \left(\frac{q_i}{a^3} r_{ij} \cdot i_1 + \frac{q_i}{r_{ij}^2} \cdot i_2 \right) ar_{ij} p_{ij} (\mathbf{X}_i - \mathbf{X}_j) \quad \begin{cases} j = 1, 2, \ldots, N \\ i_1 = 1,\ i_2 = 0 \Longleftrightarrow r_{ij} < a \\ i_1 = 0,\ i_2 = 1 \Longleftrightarrow r_{ij} \geq a \end{cases} \tag{3.3}$$

the separation distance between two charged particles defined as

$$r_{ij} = \frac{\|\mathbf{X}_i - \mathbf{X}_j\|}{\|(\mathbf{X}_i + \mathbf{X}_j)/2 - \mathbf{X}_{\text{best}}\| + \varepsilon}, \tag{3.4}$$

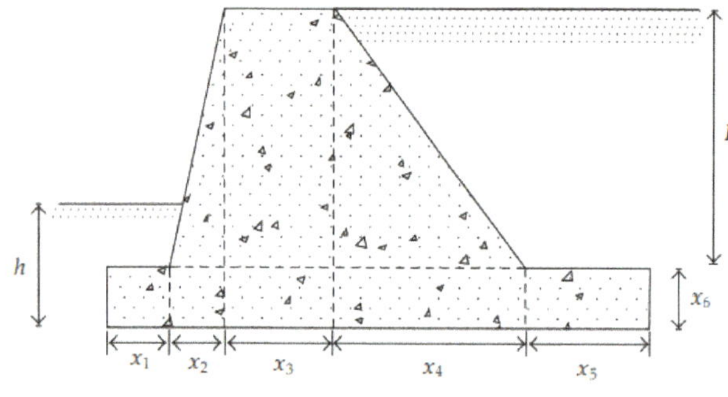

Figure 2: The design variables.

where ε is a small positive number to avoid singularity. The initial positions of CPs are determined randomly in the search space and the initial velocities of charged particles are assumed to be zero. P_{ij} determines the probability of moving each CP toward the others as

$$p_{ij} = \begin{cases} 1 & \frac{\text{fit}(i) - \text{fitbest}}{\text{fit}(j) - \text{fit}(i)} > \text{rand} \vee \text{fit}(j) < \text{fit}(i). \\ 0 & \text{otherwise.} \end{cases} \tag{3.5}$$

The resultant forces and the laws of the motion determine the new location of the CPs. At this stage, each CP moves towards its new position under the action of the resultant forces and its previous velocity as

$$\mathbf{X}_{j,\text{new}} = \text{rand}_{j1} \cdot k_a \cdot \frac{\mathbf{F}_j}{m_j} \cdot \Delta t^2 + \text{rand}_{j2} \cdot k_v \cdot \mathbf{V}_{j,\text{old}} \cdot \Delta t + \mathbf{X}_{j,\text{old}},$$

$$\mathbf{V}_{j,\text{new}} = \frac{\mathbf{X}_{j,\text{new}} - \mathbf{X}_{j,\text{old}}}{\Delta t}, \tag{3.6}$$

where k_a is the acceleration coefficient; k_v is the velocity coefficient to control the influence of the previous velocity; rand_{j1} and rand_{j2} are two random numbers uniformly distributed in the range 0, 1. If each CP moves out of the search space, its position is corrected using the harmony search-based handling approach as described in [14]. In addition, to save the best design, a memory charged memory is utilized. The flowchart of the CSS algorithm is shown in Figure 3. 4.

NUMERICAL EXAMPLE

In this section, an example is optimized with the proposed method. The final result is compared to the solution of the particle swarm optimization (PSO), big bang-big crunch algorithm BB-BC, and heuristic big bang-big crunch (HBB-BC) [15] methods to demonstrate the efficiency of the present approach. For the example presented in this paper, the CSS algorithm parameters were set as follows: $k_a = 2.0$, $k_v = 1.5$, the number of agents is taken as 20, and the maximum number of searches is set to 500. The algorithms are coded in Matlab and in order to handle the constraints, a penalty approach is utilized. If the constraints are between the allowable limits, the penalty is zero; otherwise, the amount of penalty is obtained by dividing the violation of allowable limit to the limit itself.

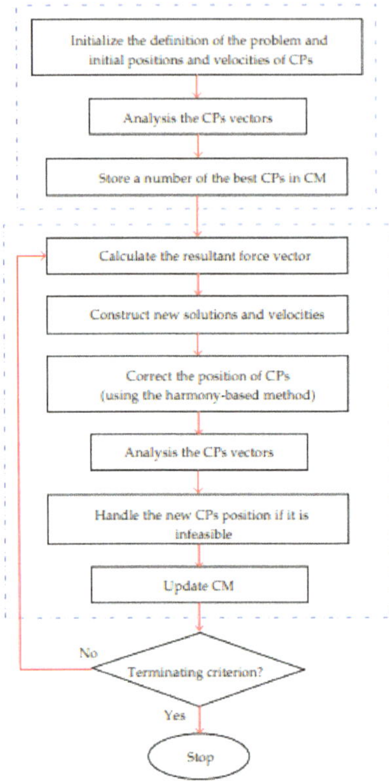

Figure 3: The flowchart for the CSS algorithm.

The problem is the optimum seismic design of a wall with H = 5.5 m and h = 1 m. The backfill has shear strength parameters of $c' = 0$, $\phi' = 30°$, and $\gamma = 16\,\mathrm{kN/m^3}$. The wall is founded on a soil with c' equals zero, $\phi' = 38°$, and $\gamma = 18.5\,\mathrm{kN/m^3}$. The horizontal and vertical ground acceleration coefficient K_h and K_v is 0.35 and 0.0. Also the material's density is 24 kN/m concrete wall. In this example, the angle of wall friction is 15◦ and the inclination of ground surface behind wall to horizontal is zero.

The results of the seismic design optimization process for the CSS algorithm and the PSO, BB-BC, and HBB-BC are summarized in Table 1. As shown in this table, the result for the CSS algorithm is 322.293 kN, which is lighter than the result of the PSO, standard BBBC, and HBB-BC algorithm. In addition, the average weight of 20 different runs for the CSS algorithm is 2.3%, 4.8%, and 6.1% lighter than the average results of the HBB-BC, BB-BC, and PSO algorithms, respectively. Comparing these results shows that the new algorithm not only improves the reliability property due to decrease in the mean of results but also enhances the quality of the results due to the decrease in the best of results. The convergence history for the CSS gravity retaining wall design is shown in Figure 4.

Table 1: The optimum seismic designs comparison for the gravity retaining wall

Design variable	Optimal values (m)			
	CSS	HBB-BC [15]	BB-BC [15]	PSO [15]
x_1	1.137	1.734	0.903	0.566
x_2	0.440	0.553	0.650	2.000
x_3	0.353	0.467	0.689	0.200
x_4	3.200	3.014	2.691	2.000
x_5	2.261	0.719	0.515	0.645
x_6	0.200	0.200	0.200	0.200
Best weight (kN)	322.293	328.297	337.607	344.700
Average weight (kN)	329.893	337.860	345.652	351.189
Number of analysis	4,400	10,000	10,000	10,000
FS_o	2.152	2.504	2.003	2.097
FS_s	1.200	1.200	1.200	1.200
FS_b	7.154	7.459	5.921	7.182

Among the design constraints, the safety factor of sliding is the active one and almost for all design of different studied algorithms, it is the most important while the factor of safety against bearing capacity is not active and it will not affect the optimum design. Any optimum design problem involves a design vector and a set of problem parameters. In many cases, we would be interested in knowing the sensitivities or derivatives of the optimum design variables and objective function with respect to the problem parameters because this is very useful to the designer, to know which data values are more influential on the design. Sensitivity of optimal responses to these parameters is one of the important issues in the optimum design of retaining walls. Here, using sensitivity analysis, the effect of changes on the safety factor for sliding on optimum weight of a wall was studied. The factor of safety for sliding of the wall is defined as the resisting forces divided by the driving force, or

$$FS_s = \frac{\text{Sliding resistance force} + \text{Allowable passive resultant force}}{\text{Active earth pressure resultant force}}. \tag{4.1}$$

If the wall is found to be unsafe against sliding, shear key below the base is provided. Such a key develops passive pressure which resists completely the sliding tendency of the wall. The customary minimum safety factor against sliding is 1.2, with some agencies requiring more. In the determination of the, FS_s, the effect of passive lateral earth pressure resistance in front of a wall footing or a wall footing key will only be considered when competent soil or rock exists which will not be removed or eroded during the structure life.

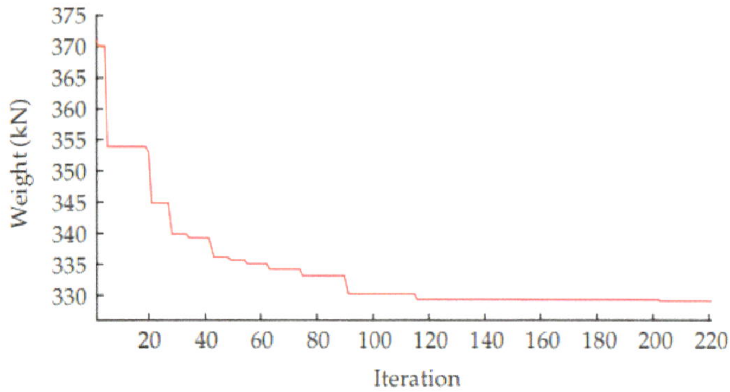

Figure 4: The convergence history of CSS algorithm average of 20 different runs.

Not more than 50 percent of the available passive lateral earth pressure will be considered in determining the FSs. In Figure 5, optimum weight variation against safety factor of sliding is depicted. It is interesting to emphasis that a small coefficient for FS_s = 1.2 causes an average decrease in cost of 43% as compared to a coefficient for FS_s = 2.

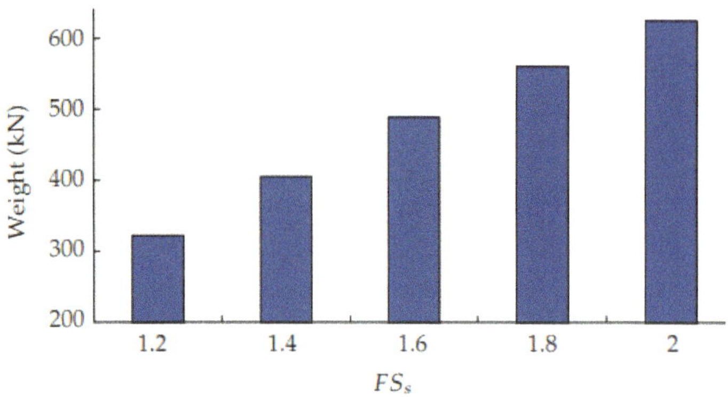

Figure 5: Weight variation against different FS_s.

CONCLUDING REMARKS

Determining optimum weight and sensitivity analysis of gravity retaining walls subject to seismic loading is presented in detail, using the CSS algorithm. This algorithm contains three levels: initialization, search, and controlling the terminating criterion. In the initialization level, the parameters of the CSS algorithm, the primary location of the CPs, and their initial velocities are defined. Also in this level, a memory to store a number of the best CPs is introduced. The search level starts after the initialization level, where each CP moves toward the others considering the probability function, the magnitude of the attracting force vector, and the previous velocities. The moving process is defined in a way that it not only can perform more investigation in the search space, but also can improve the results. To fulfill this goal, some laws of physics containing the Coulomb and Gauss laws, and the governing laws of motion from Newtonian mechanics are utilized. The last level consists of controlling the termination.

Comparing the results of the retaining wall designs obtained by other meta-heuristic algorithms such as the PSO and the BB-BC shows a good balance between the exploration and exploitation abilities of the CSS; hence, its superior performance becomes evident. Both CSS and PSO are population-based algorithms in which the position of each agent is obtained by adding the agent's movement to its previous position; however, the movement strategies are different. The PSO algorithm utilizes a velocity term being a combination of the previous velocity movement in the direction of the local best, and movement in the direction of the global best, while the CSS approach uses the governing laws from electrical physics and the governing laws of motion from the Newtonian mechanics to determine the amount and the direction of a charged particle' movement. The potency of the PSO is summarized to find the direction of an agent's movement and therefore determining the acceleration constants becomes important. Similarly in the CSS method, updating is performed by considering the quality of the solutions and the separation distances between CPs. Therefore, not only the directions but also the amounts of movements are determined.

Also a sensitivity analysis is performed for the optimum seismic design of gravity retaining wall parameters using the CSS algorithm in which the safety factor for sliding is concerned. The results related to the influence of the safety factors of sliding show that as expected, a large safety factor causes a costly wall compared to a small one.

REFERENCES

1. R. Roy, S. Hinduja, and R. Teti, "Recent advances in engineering design optimisation: challenges and future trends," CIRP Annals, vol. 57, no. 2, pp. 697–715, 2008. ·
2. M. Li, S. C. Lim, and S. Chen, "Exact solution of impulse response to a class of fractional oscillators and its stability," Mathematical Problems in Engineering, vol. 2011, Article ID 657839, 9 pages, 2011.
3. S. Chen, Y. Wang, and C. Cattani, "Key issues in modeling of complex 3D structures from video sequences," Mathematical Problems in Engineering, vol. 2012, Article ID 856523, 17 pages, 2012.
4. A. H. Gandomi, X.-S. Yang, and A. H. Alavi, "Mixed variable structural optimization using firefly algorithm," Computers and Structers, vol. 89, no. 23-24, pp. 2325–2336, 2011.
5. A. H. Gandomi, X. -S. Yang, and A. H. Alavi, "Cuckoo search algorithm: a metaheuristic approach to solve structural optimization problems," Engineering Computations. In press.
6. A. H. Gandomi and A. H. Alavi, "Krill Herd: A new bio-inspired optimization algorithm,"Communications in Nonlinear Science and Numerical Simulation, vol. 17, no. 12, pp. 4831–4845, 2012.

7. S. Chen, Y. Zheng, C. Cattani, and W. Wang, "Modeling of biological intelligence for SCM system optimization," Computational and Mathematical Methods in Medicine, vol. 2012, Article ID 769702, 30 pages, 2012.

8. S. Okabe, "General theory of earth pressure," Japanese Society of Civil Engineers, vol. 2, no. 1, 1926.

9. N. Mononobe, "Earthquake proof construction of masonry dams," in Proceedings of the World Engineering Congress, pp. 275–293, Tokyo, Japan, 1929.

10. P. N. Psarropoulos, G. Klonaris, and G. Gazetas, "Seismic earth pressures on rigid and flexible retaining walls," Soil Dynamics and Earthquake Engineering, vol. 25, no. 7–10, pp. 795–809, 2005.

11. Y. Yeung and K. Ho, Gravity Retaining Walls Subject to Seismic Loading, Civil Engineering Department, Hong Kong, China, 1994.

12. A. Kaveh and A. Shakouri Mahmud Abadi, "Harmony search based algorithms for the optimum cost design of reinforced concrete cantilever retaining walls," International Journal of Civil Engineering, vol. 9, no. 1, pp. 1–8, 2011.

13. A. C. Trandafir, T. Kamai, R. C. Sidle, and M. Popescu, "Seismic retrofit of gravity retaining walls for residential fills using ground anchors," Geotechnical and Geological Engineering, vol. 25, no. 6, pp. 679–691, 2007.

14. A. Kaveh and S. Talatahari, "Optimal design of skeletal structures via the charged system search algorithm," Structural and Multidisciplinary Optimization, vol. 41, no. 6, pp. 893–911, 2010.

15. A. Kaveh, S. Talatahari, and R. Sheikholeslami, "Optimum seismic design of gravity retaining walls using the heuristic big bang-big crunch algorithm," in Proceedings of the 2nd International Conference on Soft Computing Technology in Civil, Structural and Environmental Engineering, B. H. V. Topping and Y. Tsompanakis, Eds., Paper 4, Civil-Comp Press, 2011.

9
CHAPTER

ACTIVE THRUST ON AN INCLINED RETAINING WALL WITH INCLINED COHESIONLESS BACKFILL DUE TO SURCHARGE EFFECT

D. M. Dewaikar, S. R. Pandey, and Jagabandhu Dixit

INTRODUCTION

Active earth pressure evaluation is required for the design of geotechnical structures such as retaining walls, sheet piles, basements, and tunnels. Active thrust acting on a retaining wall is dependent of many parameters. The theories proposed by Coulomb [1] and Rankine [2] remain the fundamental approaches to analyze the active earth pressures. Coulomb [1] studied the earth pressure problems using the limit equilibrium method considering a triangular wedge of backfill behind a rough retaining wall with a plane failure surface and this theory is well verified for the frictional soil in active state. The point of application of active thrust is assumed at a distance one-third of the height of the wall from its base and independent of different parameters such as soil friction angle, ϕ, angle of wall friction, δ, backfill angle, β, and wall inclination angle. Coulomb's [1] approach does not use moment equilibrium equation for the analysis, since the distribution and point of application of reaction on the failure plane are unknown. If the distribution and point of application of soil reaction on the failure plane are known, then the point of application of active thrust can be determined using moment equilibrium equation. The limit equilibrium method assuming appropriate failure surface is most frequently used to analyze static earth pressure.

A failure mechanism was proposed by Terzaghi [3] that assumes a log-spiral failure surface originating from the base of the wall, followed by a tangent, that meets the ground surface at an angle corresponding to Rankine's [2] active earth pressure. Several experimental, analytical, and numerical studies were performed to evaluate the earth pressures on the retaining walls [4–14]. Some rigorous approaches such as finite element methods were attempted to determine the distribution of earth pressures on the retaining wall. Others prepared the tables for the calculation earth pressures based

Citation: D. M. Dewaikar, S. R. Pandey, and Jagabandhu Dixit, "Active Thrust on an Inclined Retaining Wall with Inclined Cohesionless Backfill due to Surcharge Effect," ISRN Soil Science, vol. 2012, Article ID 750386, 8 pages, 2012. doi:10.5402/2012/750386.

on log-spiral failure surface [15]. Active earth pressure coefficients for cohesionless soil were also computed using the method of slices by considering the soil mass in the state of limit equilibrium [16, 17]. The charts were prepared for the calculation earth pressures based on log-spiral failure surface [18]. Discrete element analysis is used to evaluate active and passive pressure distributions on the retaining wall [19].

Recently the active earth pressure coefficients were computed based on the lower bound theorem of plasticity [20]. Upper-bound theorem of limit analysis was used to evaluate earth pressure coefficients due to soil weight, vertical surcharge loading, and soil cohesion for the case of an inclined wall and sloping cohesionless backfill [21]. Kötter's [22] equation was suitably used to determine the point of application of active thrust by taking moments of all the forces and reaction about the base of the retaining wall [23]. A method was proposed by Kame et al. [24] to determine the active earth pressure and its point of application on a vertical wall retaining horizontal cohesionless backfill based on log-spiral failure mechanism coupled with Kötter's equation. Dewaikar et al. [25] used Kötter's equation to evaluate the magnitude of active thrust on an inclined wall retaining horizontal cohesionless backfill with uniform surcharge and to find out the point of application of reaction on the failure surface.

The objective of this study is to present a mathematical approach with a high degree of simplicity for the evaluation of the active thrust and its point of application. The analysis considers the free body diagram of the retaining wall in a very simple and analytical procedure and is easier for engineering application. The validity of the proposed approach is checked with previous available results.

ANALYSIS OF ACTIVE THRUST

The active thrust, Pa, is determined from the conditions of force equilibrium and its point of application is determined from the conditions of moment equilibrium. Figure 1 shows an inclined retaining wall, with inclined cohesionless backfill subjected to a uniform surcharge intensity of q KN/m^2. A trial plane failure surface [1] is considered which meets the ground surface at an angle, α with the horizontal. The trial wedge ABC is in equilibrium under the effect of three different forces: (1) the equivalent force of the surcharge, W (2)the soil reaction, R along the face trial failure BC, at an angle ϕ to the normal on BC and (3) the active thrust Pa, at an angle δ to the normal on the back.

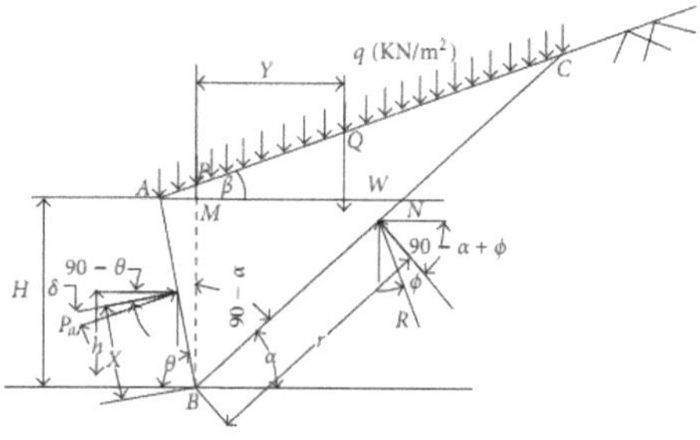

Figure 1: Free body diagram of the trial failure wedge ABC.

The symbols used in Figure 1 are defined as follows. Pa: active thrust. W: equivalent force of the surcharge. R: soil reaction on the failure wedge. H: height of the retaining wall. h: height of point of application of active thrust from the wall base. θ: inclination of the retaining wall with the horizontal. δ: friction angle between the wall and soil backfill. ϕ: soil friction angle. α: inclination of the trial failure plane with the horizontal. q: intensity of surcharge in kN/m^2.

Equating all the forces in the vertical and horizontal direction, the following force equilibrium conditions are obtained.

Horizontal force equilibrium:

$$P_a \sin(\theta - \delta) = R\sin(\alpha - \phi) \tag{1}$$

from which R is obtained as

$$R = \frac{P_a \sin(\theta - \delta)}{\sin(\alpha - \phi)}. \tag{2}$$

Vertical force equilibrium:

$$P_a \cos(\theta - \delta) + R\cos(\alpha - \phi) = AC \cdot q \tag{3}$$

substituting the value of R from (2) into (3)

$$P_a \cos(\theta - \delta) + \frac{P_a \sin(\theta - \delta)}{\sin(\alpha - \phi)} \cos(\alpha - \phi) = AC \cdot q, \tag{4}$$

$$P_a \cos(\theta - \delta) + P_a \sin(\theta - \delta)\cot(\alpha - \phi) = AC \cdot q. \tag{5}$$

From (5), P_a is obtained as

$$P_a = \frac{AC \cdot q}{\cos(\theta - \delta) + [\sin(\theta - \delta)\cot(\alpha - \phi)]}. \tag{6}$$

Now, applying sine rule to the triangle CAN,

$$\frac{AC}{\sin(180 - \alpha)} = \frac{AN}{\sin(\alpha - \beta)} \tag{7}$$

Or

$$AC = \frac{AN \cdot \sin(180 - \alpha)}{\sin(\alpha - \beta)}. \tag{8}$$

Referring to Figure 1,

$$AN = H(\cot\theta + \cot\alpha). \tag{9}$$

Substituting the value of AN in (8), AC is obtained as,

$$AC = \frac{H(\cot\alpha + \cot\theta)\sin\alpha}{\sin(\alpha - \beta)}. \tag{10}$$

Substituting the value of AC in (6), the magnitude of active thrust Pa is expressed as

$$P_a = \frac{qH(\cot\alpha + \cot\theta)\sin\alpha}{\sin(\alpha - \beta)[\cos(\theta - \delta) + \sin(\theta - \delta)\cot(\alpha - \phi)]}. \tag{11}$$

The maximum value of active thrust (P_a) is obtained when the inclination of the failure plane, BC, with the horizontal reaches the critical value, α_{cr}.

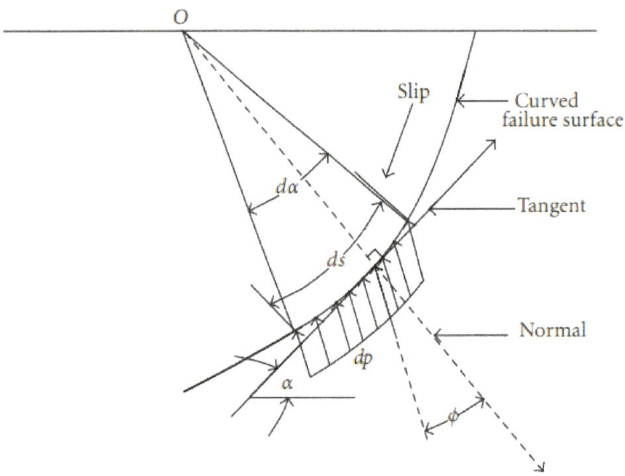

Figure 2: Reactive pressure distribution along the failure surface.

EVALUATION OF SOIL REACTION ON THE FAILURE SURFACE

In a cohesionless soil medium with active state of equilibrium under plane strain condition, Kötter's [22] equation is given as (Figure 2)

$$\frac{dp}{ds} - 2p\tan\phi\frac{d\alpha}{ds} = \gamma\sin(\alpha - \phi)s, \tag{12}$$

where dp: differential reactive pressure at a point on the failure surface, ds: elemental length of the failure surface, α: angle made by the tangent at the point of interest with the horizontal, ϕ: soil friction angle and, γ: soil unit weight.

In the present analysis, only surcharge effect is taken into consideration, and therefore γ becomes zero. With this consideration (12) can be written as

$$\frac{dp}{d\alpha} = 2p\tan\phi \tag{13}$$

Or

$$dp = 2p\tan\phi d\alpha. \tag{14}$$

For a plane failure surface, dα is zero and the corresponding solution for the reactive pressure, p, is obtained as

$$p = \text{constant}. \tag{15}$$

The above solution indicates that soil reactive pressure (p) is uniformly distributed along the failure plane, BC. Therefore, the resultant soil reaction, R, acts at the mid-point of the failure plane, BC. The magnitude of soil reaction is computed after knowing the critical value, α_{cr}, of the angle, α. Substituting the value of P_a from (11) in to (2), the soil reaction R on the failure surface is obtained as

$$R = \frac{qH(\cot\alpha_{cr} + \cot\theta)\sin(\theta-\delta)\sin\alpha_{cr}}{\sin(\alpha_{cr}-\phi)\sin(\alpha_{cr}-\beta)} \times \frac{1}{[\cos(\theta-\delta)+\sin(\theta-\delta)\cot(\alpha_{cr}-\phi)]}. \tag{16}$$

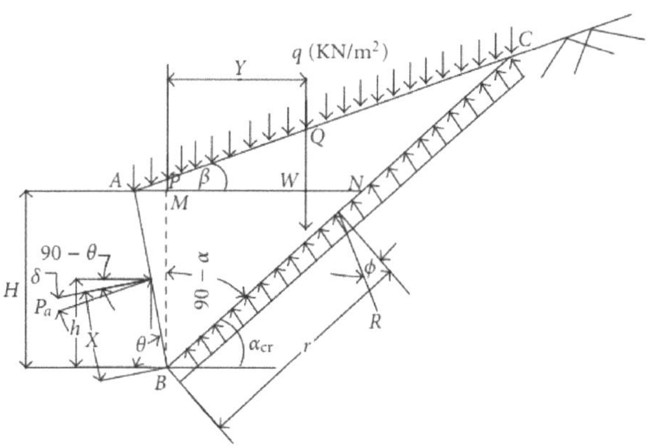

Figure 3: Free body diagram of the failure wedge ABC.

POINT OF APPLICATION OF THE ACTIVE THRUST

After obtaining the value of P_a, the condition of moment equilibrium is applied. Figure 3 shows the free body diagram of the failure wedge ABC. The equivalent force of surcharge on the trial wedge, ABC, is AC·q, which acts at the midpoint of AC, that is, at a distance Y from the vertical line passing through the base the wall. As the distribution of reaction (R) is uniform along the failure surface, it acts at the mid-point of the failure plane BC. The moments of all forces and reactions are taken about the base of the wall, at point B.

Moment equilibrium condition:

$$P_a \cos\delta \cdot X + qAC \cdot Y = R\cos\phi \frac{BC}{2}. \tag{17}$$

The surcharge acts at the center of AC with the distance Y given as,

$$Y = PQ\cos\beta, \tag{18}$$

Where

$$PQ = \frac{AC}{2} - AP. \tag{19}$$

From triangle AMP,

$$AP = \frac{H\cot\theta}{\cos\beta}. \tag{20}$$

Now, substituting the value of AC from (10) and AP from (20) in (19) (Figure 3), PQ is obtained as

$$PQ = \frac{H(\cot\theta + \cot\alpha_{cr})\sin\alpha_{cr}}{2\sin(\alpha_{cr} - \beta)} - \frac{H\cot\theta}{\cos\beta}. \tag{21}$$

Table 1: Variation of the height of point of application of active thrust, Hr, with angle of wall back (θ), angle of soil friction (φ), soil and wall friction (δ) and backfill slope (β).

Angle of wall back, θ (degrees)	Angle of soil friction, ϕ (degrees)	Angle of wall friction, δ (degrees) ($\delta = 2/3\phi$)	Angle of backfill slope, β (degrees) $\beta/\phi = (0.4, 0.6, 0.8)$	H_r ($=h/H$)
			16	0.336
	40	26.667	24	0.397
			32	0.511
			14	0.388
	35	23.333	21	0.451
80			28	0.566
			12	0.446
	30	20	18	0.510
			24	0.622
			10	0.512
	25	16.667	15	0.575
			20	0.682
			16	0.657
	40	26.667	24	0.708
			32	0.814
			14	0.625
	35	23.333	21	0.68
70			28	0.775
			12	0.605
	30	20	18	0.648
			24	0.729
			10	0.584
	25	16.667	15	0.625
			20	0.692

Substituting the value of PQ from (21) into (18), the point of application of equivalent surcharge force from the axis BM (Figure 3) is computed as

$$Y = \cos\beta\left[\frac{H(\cot\theta+\cot\alpha_{cr})\sin\alpha_{cr}}{2\sin(\alpha_{cr}-\beta)}-\frac{H\cot\theta}{\cos\beta}\right]. \tag{22}$$

The soil reaction R acts at the midpoint of the failure plane BC, with the distance, r (Figure 3), computed as

$$r = \frac{BC}{2}. \tag{23}$$

Referring to Figure 3, BC is obtained as

$$BC = \frac{H\sin(\theta+\beta)}{\sin(\alpha_{cr}-\beta)\sin\theta}. \tag{24}$$

Substituting the value of BC from (24) and Y from (22) into (17), the distance X is obtained:

$$\begin{aligned} &P_a\cos\delta\cdot X + q\cdot AC\cos\beta \\ &\quad\times\left[\frac{H(\cot\theta+\cot\alpha_{cr})\sin\alpha_{cr}}{2\sin(\alpha_{cr}-\beta)}-\frac{H\cot\theta}{\cos\beta}\right] \\ &\quad= R\cos\phi\frac{H\sin(\theta+\beta)}{2\sin(\alpha_{cr}-\beta)\sin\theta} \end{aligned} \tag{25}$$

Or

$$\begin{aligned} X = &\frac{R\cos\phi\dfrac{H\sin(\theta+\beta)}{2\sin(\alpha_{cr}-\beta)\sin\theta}}{P_a\cos\delta} \\ &-\frac{q\cdot AC\cos\beta\left[\dfrac{H(\cot\theta+\cot\alpha_{cr})\sin\alpha_{cr}}{2\sin(\alpha_{cr}-\beta)}-\dfrac{H\cot\theta}{\cos\beta}\right]}{P_a\cos\delta}. \end{aligned} \tag{26}$$

The height, h, of point of application of P_a from the wall base is obtained as

$$h = X\sin\theta. \tag{27}$$

The coefficient of active earth pressure (K_{aq}) for the retaining wall with the surcharge effect is obtained as

$$K_{aq} = P_a/\gamma H. \tag{28}$$

RESULTS AND DISCUSSIONS

One of the aims of the proposed analysis is the determination of point of application of the active thrust in a nondimensional form with the consideration of effect of various parameters. The height, h, of point of application of the active thrust due to surcharge effect is expressed in a nondimensional form, Hr, expressed as h/H. The results are reported as shown in Table 1 along with figures that are described below.

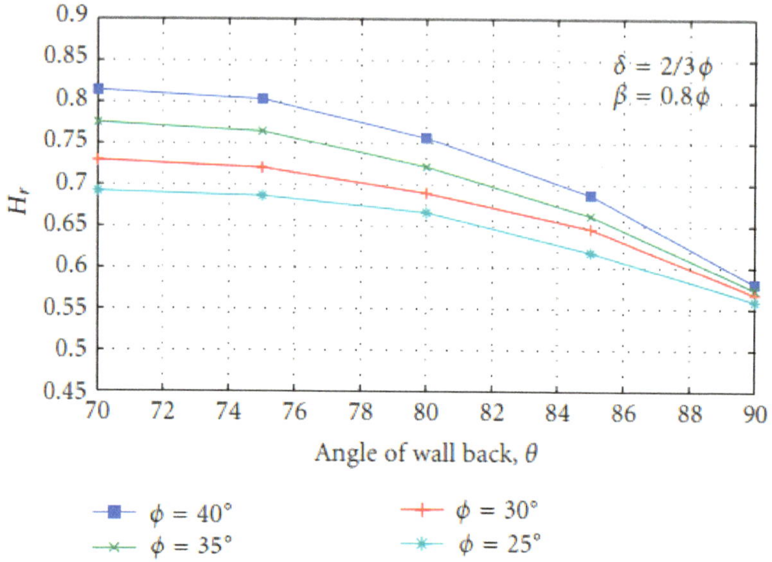

Figure 4: Variation of H_r with angle of wall back, θ, for β = 0.8φ.

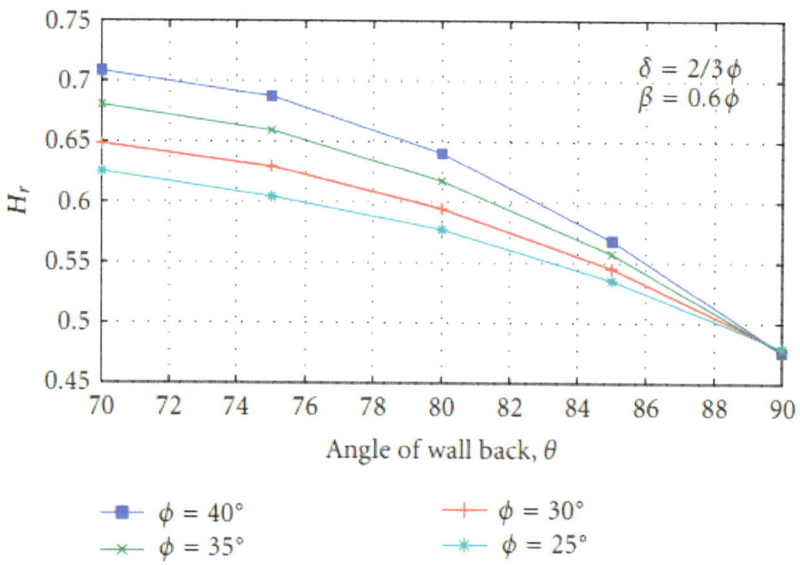

Figure 5: Variation of H_r with angle of wall back, θ, for β = 0.6φ.

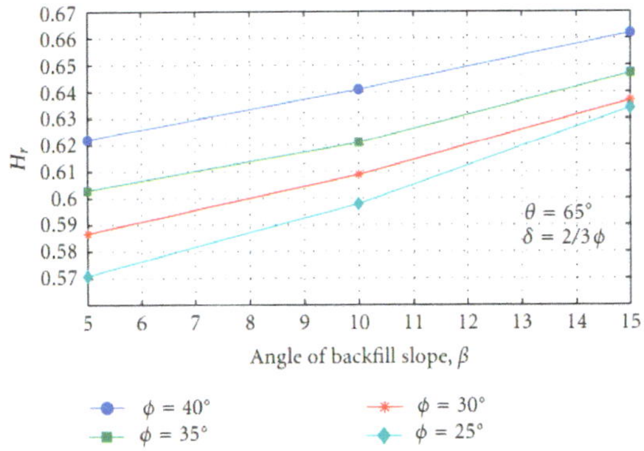

Figure 6: Variation of H_r with angle of backfill slope, β, for θ = 65°.

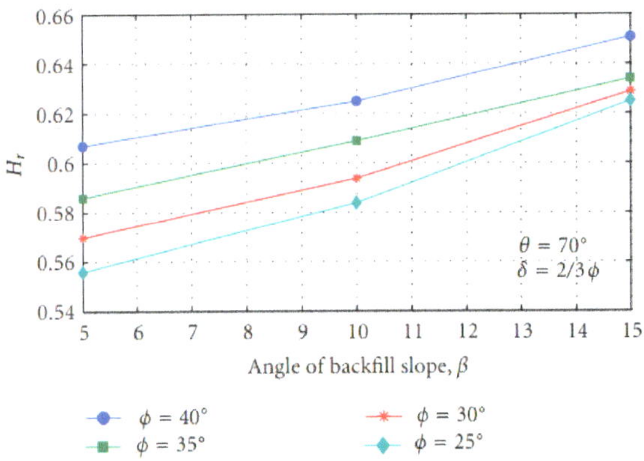

Figure 7: Variation of H_r with angle of backfill slope, β, for θ = 70°.

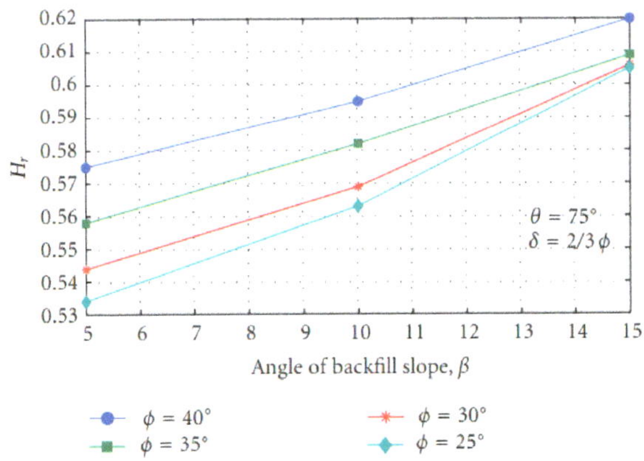

Figure 8: Variation of H_r with angle of backfill slope, β, for θ = 75°.

In Figures 4 and 5, variation of H_r with the angle of wall back, θ, is shown for δ = 2/3φ and β = 0.8φ and 0.6φ respectively. It is seen that H_r, decreases with increasing angle of wall back and increases with increasing soil friction angle, φ.

In Figures 6, 7, and 8, variation of H_r with the angle of backfill slope, β, with δ=2/3ϕ is shown for θ = 65°, 70°, and 75°, respectively. It is seen that H_r, increases with increasing β, with higher values for higher soil friction angle, ϕ.

In Figure 9, variation of H_r with the angle of wall friction, δ, is shown for β= 16° and ϕ = 40°. It is seen that H_r increases with δ, with higher values for higher β.

In Figure 10 is shown the variation of H_r for a vertical wall (θ = 90°) and δ=2/3ϕ. It is again seen that H_r increases with increase of β.

In Figure 11, variation of H_r with angle of soil friction, ϕ, is shown for a retaining wall with angle of wall back,θ = 90°, backfill slope, β = 0°, and angle of wall friction, δ = 0°. It is interesting to note that H_r is a constant value of 0.5 for all values of angle of soil friction, ϕ. Hence, the point of application of active thrust acts at the mid-height of a smooth vertical retaining wall with horizontal backfill.

Table 2: Comparison of K_{aq} with the results for the angle of wall back of θ = 80°

Soil friction angle, ϕ (degrees)	Angle of backfill slope, β (degrees)	Angle of wall friction, δ (degrees)	P_a (KN/m)	K_{aq} Present study	K_{aq} Soubra and Macuh [21]	% difference
40	16	26.667	0.4758	0.336	0.331	1.51
40	24	26.667	0.561	0.396	0.382	3.66
40	32	26.667	0.723	0.511	0.475	7.57
35	14	23.333	0.549	0.388	0.383	1.35
35	21	23.333	0.638	0.451	0.439	2.73
35	28	23.333	0.8	0.565	0.538	5.00
30	12	20	0.631	0.446	0.44	1.36
30	18	20	0.721	0.509	0.499	2.00
30	24	20	0.88	0.622	0.602	3.32
25	10	16.667	0.724	0.512	0.504	1.58
25	15	16.667	0.813	0.574	0.566	1.41
25	20	16.667	0.964	0.681	0.668	1.90

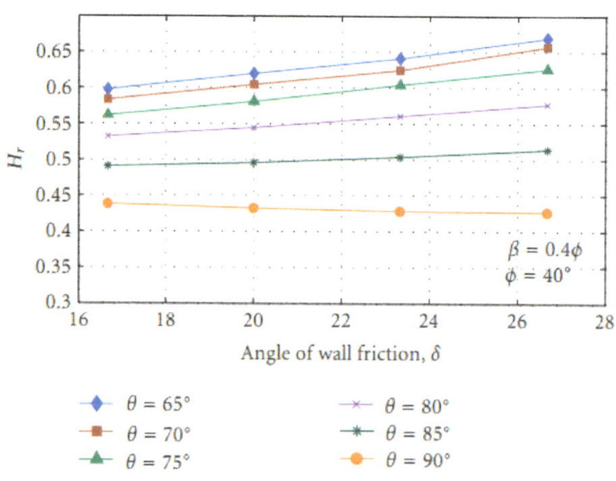

Figure 9: Variation of H_r with angle of wall friction, δ, for φ = 40°.

The other aim of the proposed analysis is to compute the values of the active earth pressure coefficient, Ka_q, for the various combinations of parameters considered in the analysis and compare the results with the available solution reported by Soubra and Macuh [21]. The results are presented through figures. In Figures 12, 13, and 14, the variation of active earth pressure coefficient, K_{aq} with angle of soil friction, φ is shown for β = 0.4φ, 0.6φ, and 0.8φ, respectively, for various values of angle of wall back, θ and δ = 2/3φ. It is seen that K_{aq} values decrease with increasing angle of soil friction, φ. It is further seen that the values show an increasing trend with decreasing values of angle of wall back, θ.

In Figure 15 and Table 2 is shown the comparison of active earth pressure coefficients obtained from the proposed method and the method proposed by Soubra and Macuh [21]. The Kaq values obtained from the proposed method are higher with the maximum difference of 7.57% for the angle of wall back of θ = 80°, φ ≤ 40°, and δ = 2/3φ as shown in Table 2. For lower values of soil friction angle, φ, and angle of backfill slope, β, the difference between the two results is less, showing an increasing trend with increasing angle of soil friction, φ, and backfill slope, β.

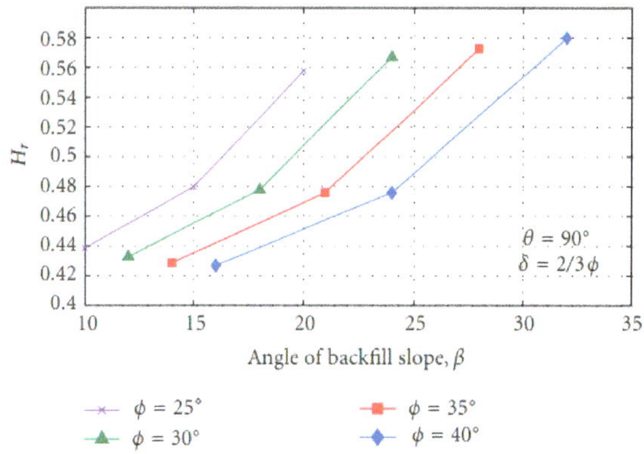

Figure 10: Variation of H_r with angle of backfill slope, β, for θ = 90°.

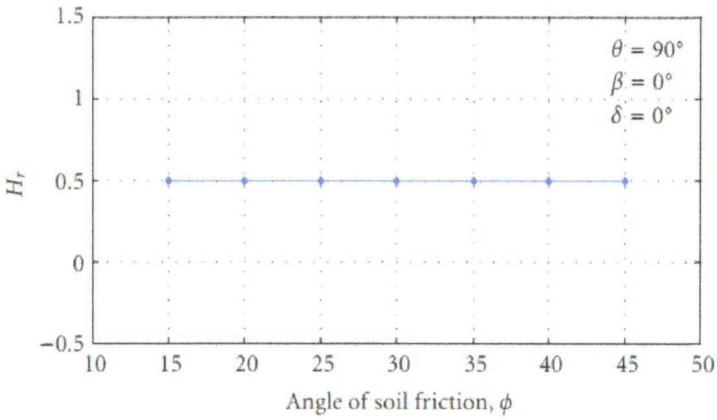

Figure 11: Variation of H_r with angle of soil friction, φ.

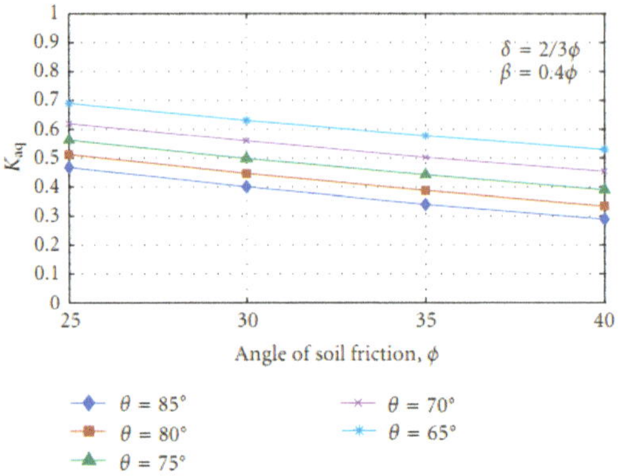

Figure 12: Variation of K_{aq} with angle of soil friction, φ, for β = 0.4φ.

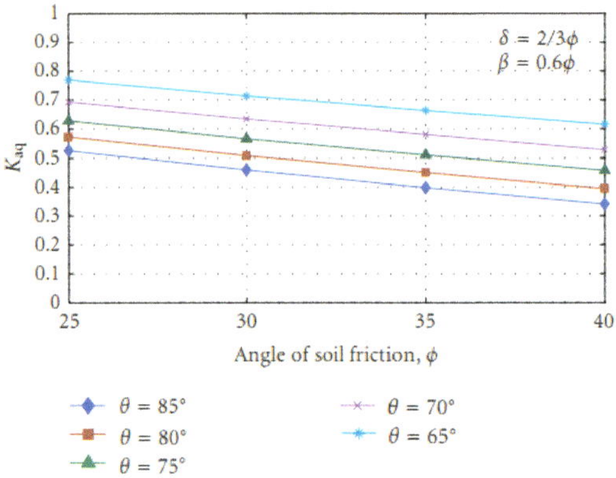

Figure 13: Variation of K_{aq} with angle of soil friction, φ, for β = 0.6φ.

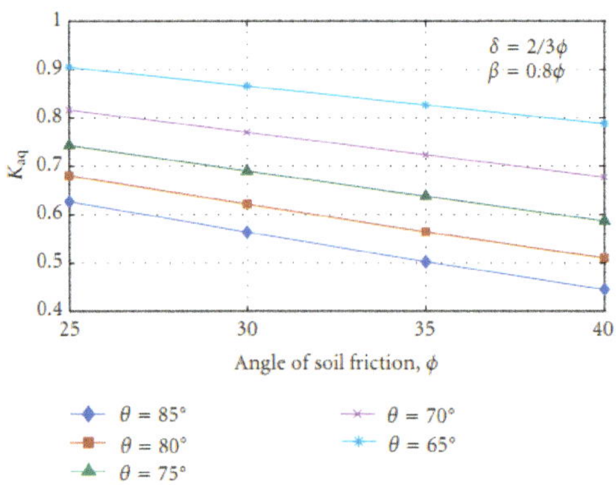

Figure 14: Variation of K_{aq} with angle of soil friction,φ, for β = 0.8φ.

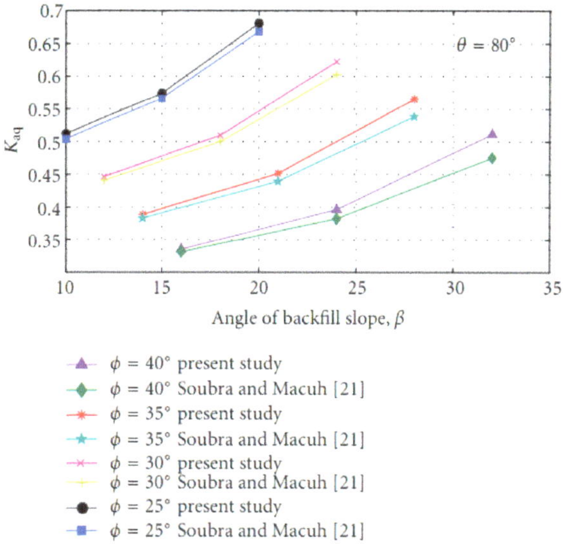

Figure 15: Comparison of variation of K_{aq} with angle of backfill slope, β, evaluated in present study with filed results.

CONCLUSIONS

The complete solution to a retaining wall problem is obtained only when the point of application of active thrust is known. Kötter's [22] equation lends itself as a powerful tool in the proposed analysis to determine the reactive pressure distribution on the failure plane. The moment equilibrium condition is used effectively to compute the point of application of the active thrust. From the proposed analysis it is seen that the point of application of the active thrust depends upon a number of factors such as angle of soil friction, ϕ, angle of wall friction, δ, angle of wall back, θ, and inclination of backfill, β. It shows a wide variation in the range, 0.427 to 0.814. Only for the case of a smooth vertical wall retaining horizontal backfill, the point of application of the active thrust is at the wall mid-height. The present approach is easy to adapt for retaining walls and the results obtained based on this analysis are in fairly good agreement with other results.

REFERENCES

1. C. A. Coulomb, "Essaisurune application des règles des maximis et minimis à quelquesproblèmes de statiquerelatifs à l'architecture," Mémoires de Mathématiques et de Physique Présentés à l'Académie Royale des Sciences par Divers Savants, vol. 7, pp. 343–382, 1776.
2. W. J. M. Rankine, "On the stability of loose earth," Philosophical Transactions of the Royal Society, vol. 147, 1857.
3. K. Terzaghi, Theoretical Soil Mechanics, John Wiley & Sons, New York, NY, USA, 1943.

4. J. B. Hansen, Earth Pressure Calculation, Danish Geotechnical Press, Copenhagen, Denmark, 1953.
5. V. V. Sokolovski, Statics of Soil Media, Butterworth, London, UK, 1960.
6. Z. V. Tsagareli, "Experimental investigation of the pressure of a loose medium on retaining walls with a vertical back face and horizontal backfill surface," Soil Mechanics and Foundation Engineering, vol. 2, no. 4, pp. 197–200, 1967.
7. M. Matsuo, S. Kenmochi, and H. Yagi, "Experimental study on earth pressure of retaining wall by field tests," Soils and Foundations, vol. 18, no. 3, pp. 27–41, 1978.
8. K. Habibagahi and A. Ghahramani, "Zero extension line theory of earth pressure," Journal of the Geotechnical Engineering Division, vol. 105, no. 14702, pp. 881–896, 1979.
9. Y. S. Fang and I. Ishibashi, "Static earth pressures with various wall movements," Journal of Geotechnical Engineering, vol. 112, no. 3, pp. 317–333, 1986.
10. D. M. Potts and A. B. Fourie, "A numerical study of the effects of wall deformation on earth pressures,"International Journal for Numerical & Analytical Methods in Geomechanics, vol. 10, no. 4, pp. 383–405, 1986.
11. E. Motta, "Generalized Coulomb active earth pressure for a distanced surcharge," Journal of Geotechnical Engineering, vol. 120, no. 6, pp. 1072–1079, 1994.
12. H. Hazarika and H. Matsuzawa, "Wall displacement modes dependent active earth pressure analyses using smeared shear band method with two bands," Computers and Geotechnics, vol. 19, no. 3, pp. 193–219, 1996.
13. V. R. Greco, "Active earth thrust by backfills subject to a line surchange," Canadian Geotechnical Journal, vol. 42, no. 5, pp. 1255–1263, 2005.
14. Y. M. Cheng, Y. Y. Hu, and W. B. Wei, "General axisymmetric active earth pressure by method of characteristics—theory and numerical formulation," International Journal of Geomechanics, vol. 7, no. 1, pp. 1–15, 2007.
15. A. I. Caquot and J. Kerisel, Tables for the Calculation of Passive Pressure, Active Pressure, and Bearing Capacity of Foundations, Gauthier-Villars, Paris, France, 1948.
16. N. Janbu, "Earth pressure and bearing capacity calculations by generalized procedure of slices," inProceedings of the 4th International Conference on Soil Mechanics and Foundation Engineering, vol. 2, pp. 207–212, London, UK, 1957.
17. H. Rahardjo and D. G. Fredlund, "General limit equilibrium method for lateral earth force," Canadian Geotechnical Journal, vol. 21, no. 1, pp. 166–175, 1984.
18. J. Kerisel and E. Absi, Active and Passive Earth Pressure Tables, Balkema, Rotterdam, The Netherlands, 1990.
19. C. S. Chang and S. J. Chao, "Discrete element analysis for active and passive pressure distribution on retaining wall," Computers and Geotechnics, vol. 16, no. 4, pp. 291–310, 1994.

20. R. Lancellotta, "Analytical solution of passive earth pressure," Geotechnique, vol. 52, no. 8, pp. 617–619, 2002.
21. A. H. Soubra and B. Macuh, "Active and passive earth pressure coefficients by a kinematical approach,"Proceedings of the Institution of Civil Engineers, vol. 155, no. 2, pp. 119–131, 2002.
22. F. Kötter, Die Bestimmung des Drucks an gekrümmten Gleitflächen, eine Aufgabe aus der Lehre vom Erddruck, Sitzungsberichte der Akademie der Wissenschaften, Berlin, Germany, 1903.
23. D. M. Dewaikar and S. A. Halkude, "Seismic passive/active thrust on retaining wall-point of application,"Soils and Foundations, vol. 42, no. 1, pp. 9–15, 2002.
24. G. S. Kame, D. M. Dewaikar, and D. Choudhury, "Active thrust on a vertical retaining wall with cohesionless backfill," Electronic Journal of Geotechnical Engineering, vol. 15, pp. 1848–1863, 2010.
25. D. M. Dewaikar, S. R. Pandey, and J. Dixit, "Active earth pressure on an inclined wall with horizontal cohesionless backfill due to surcharge effect," Electronic Journal of Geotechnical Engineering, vol. 17, pp. 811–824, 2012.

10
CHAPTER

DYNAMIC RESPONSE OF WALL BACKFILL RETAINING SYSTEM

Sreenivas Alampalli and Ahmed-W. Elgamal

INTRODUCTION

Retaining walls are built for the purpose of supporting an adjacent soil mass or backfill. The magnitude of earthquake-induced loads on such structures depends on the backfill and foundation properties, earthquake motion characteristics, and structural wall properties. Damage to retaining wall systems due to seismic activity is often reported in the literature (Jennings, 1971; Seed and Whitman, 1970; Sitar et al., 1995; Tatsuoka et al., 1996; Whitman and Christian, 1990). A number of analytical and numerical models have been proposed to investigate the earthquake response of cantilever retaining walls (Alampalli, 1990; Whitman, 1991). In these models, earthquake response is assumed to be uniform along the wall, and such wall backfill systems are modeled under 2-dimensional (2- D) or plane-strain conditions (Arias et al., 1981; Matsuo and Ohara, 1960; Velestos et al., 1995; Wood, 1973, 1975).

A large and growing body of full-scale dynamic tests on civil engineering structures has been reported in the literature (Alampalli et al., 1995; Calciati et al., 1977; Douglas and Reid, 1982; Keightley, 1966; Paskalov et al., 1977). However, very few full-scale tests have been performed on retaining wall soil systems (Aliev et al., 1973; Amano et al., 1956; Chang et al., 1990; Fukuoka and Imamura, 1984). During a pilot full-scale test on a retaining wall system (Alampalli, 1990; Elgamal et al., 1990), it was found that modal spatial variation was displayed along the wall length as well as its height. Modes were found to bear significant similarity to those of a clamped cantilever plate rather than those of a cantilever beam (planestrain analysis). Such a 3-D response pattern was further analyzed during a more recent forced vibration frequency sweep test (Elgamal et al., 1996). In the following section, the experimental testing program is summarized. A number of salient experimental observations are presented. The recorded dynamic

Citation: Sreenivas Alampalli and Ahmed-W. Elgamal, "Dynamic Response of Wall Backfill Retaining System," Shock and Vibration, vol. 4, no. 4, pp. 251-259, 1997. doi:10.3233/SAV-1997-4405.

response is analyzed within a simple analytical and numerical framework to investigate the observed 3-D response mechanisms.

EXPERIMENTAL STUDY

The test structure was a long wall (140 ft) of varying height from 4.6 to 10 ft and a thickness of 1.3 ft (Fig. 1). This wall supported an elevated parking lot. A construction joint was located at about 75 ft from the short end of the wall (Fig. 1).

A typical test configuration is shown in Fig. 2. It consists of a dynamic signal analyzer (HP model HP3562A) to obtain real-time frequency domain representations of the recorded signals, a large 12-lb weight sledge hammer with a force transducer for excitation (PCB model 086B50), an accelerometer for response measurement (PCB model 39C), amplifiers, signal conditioners, and a microcomputer with a floppy and hard disk drive (HP model 300, series 9000). At the beginning of testing, reciprocity was verified by comparing frequency response functions (FRF) obtained through interchanging input and output locations. During a test, measurement locations were chosen to reflect the behavior of the strucutre in the modes of interest. Only the dynamic wall response perpendicular to its plane was measured (Alampalli, 1990).

Excitation hammer input was imparted at sampling points (measurement locations) and the resulting acceleration output was measured at a stationary point (Alampalli, 1990). For each measurement point, an input-output FRF was computed by the analyzer, which also performed signal digilization using appropriate antialiasing filters, such that,

$$\mathrm{FRF}(f) = G_{yx}(f)/G_{xx}(f), \tag{1}$$

where $G_{yx}(f)$ is the cross power sectrum of output and input signals, G_{xx} (f) is the auto power spectrum of the input signal, and f is the frequency variable. An exponential window (filter) was used on the structural vibration response and a force window was used on the hammer excitation. Mode shapes were extracted using the peak curve-fitting technique (Ewins, 1984). For any sampling point, the process was repeated and the average of a number of FRFs was finally stored as the input-output transfer function for this point. Inspection of the coherence function, γ (Ewins, 1984) dictated the number of FRFs to be averaged, where

$$\gamma^2(f) = G_{yx}(f)G_{xy}(f)/\big(G_{xx}(f)G_{yy}(f)\big), \tag{2}$$

in which G_{yy}(f) is the auto power spectrum of the output signal. The coherence function assures consistency of the obtained data. A coherence of zero denotes that input and output are unrelated, whereas a coherence of unity denotes that they are related with no interference or noise effects. In this study, a highly satisfactory coherence of 0.95 or greater was always achieved. A modal parameter estimation package (modal 3.0 SE), developed by Structural Measurement Systems (SMS, 1987), was employed for extraction of modal parameters (linearity and reciprocity for the test structures is assumed).

Transfer functions of 178 measurement locations along the exposed side of the wall were measured (Alampalli, 1990). From these transfer functions, resonant frequencies and associated mode shapes were obtained. A number of these mode shapes are shown in Fig. 3(a). A rough estimate of damping ratios suggested a value of about 8% for most of these modes. In a more recent test, the same structure was tested in the low frequency range (up to 17 Hz) of significance for earthquake-type excitation (Elgamal et al., 1996). A 5000-lb harmonic shaker was employed in a frequency sweep mode. The observed modal configurations [Fig. 3(b)] also clearly displayed the above mentioned 3-D response effects.

It may be noted that the resonant wall configurations (Fig. 3) mimic that of a cantilever clamped plate. Along the length, the amplitude gradually increased with the increase in free cantilever wall height (Fig. 3). Such a 3-D response pattern is not accounted for by current analysis procedures. Hence, a simple 3-D finite element model was employed to further analyze the observed resonant configurations and their potential contribution to seismic response of wall backfill soil systems.

NUMERICAL STUDY

A simple finite element model was employed to investigate the conditions associated with 3-D seismic retaining wall response. The wall was modeled using 3-D brick elements. A system of springs along the wall face represented the backfill soil. In this model, the following parameters were employed: Young's modulus = 4 x 10^8 lb/ft^2, Poisson's ratio = 0.15, density = 145 lb/ft^3, and soil spring constant = 1 x 10^5 lb/ft^3.

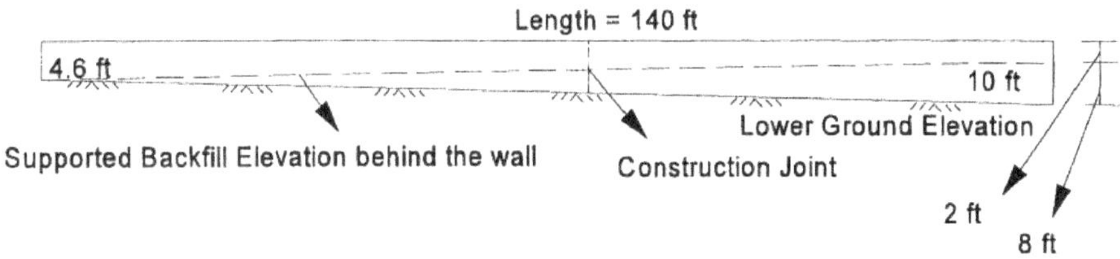

Figure 1: Schematic of tested wall (elevation view).

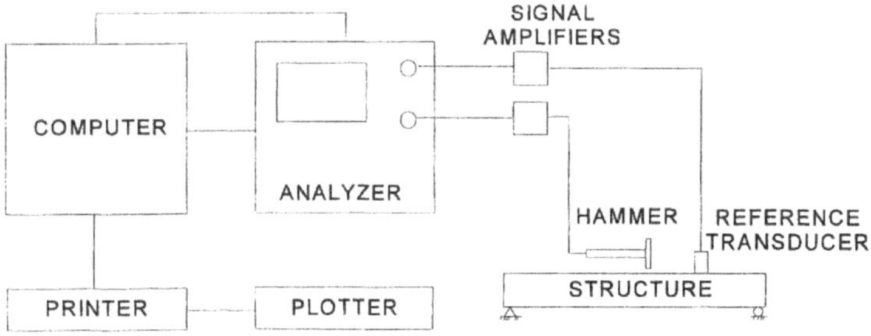

Figure 2: General test setup for impact hammer test.

Table 1: Natural Frequencies and Modal Participation Factors of 30-ft Plane-Strain Wall

Mode No.	Frequency (Hz)	Modal Participation Factor
1	5.581	0.50285
2	17.186	−0.18740
3	31.237	−0.00016
4	47.088	0.12500
5	90.638	−0.03567

The response of four long walls (fixed at the base and free along the boundaries) was studied: a wall of 30-ft height and 3-ft thickness, using plane-strain analysis; a wall of 150-ft length, 3-ft thickness, and 30- ft uniform height; a wall of 15O-ft length, 3-ft thickness, and height varying from 40 ft at one end to 20 ft at the other end; and a wall of 150-ft length, 3-ft thickness, and height varying from 40 ft at one end to 10 ft at the other end. In all cases, the retained soil was assumed to extend throughout the wall height and length. Natural frequencies and modal participation factors (perpendicular to the wall face) were computed by an appropriate eigenvalue solver (Tables 1-4). Mode shapes for this 3-D finite element model are shown in Figs. 4-6. These results reveal the following.

Table 2: Natural Frequencies and Modal Participation Factors of 3-D Wall with 30-ft Uniform Height

Mode No.	Frequency (Hz)	Modal Participation Factor*
1	5.020	1.407
2	5.083	0
3	5.305	0
4	5.793	0
5	6.703	0
6	8.190	0
7	10.311	0
8	13.084	0
9	14.852	0
10	16.466	0
11	16.469	0
12	17.238	−0.720
13	17.375	0
14	17.840	0
15	18.621	0

* Motion perpendicular to wall face.

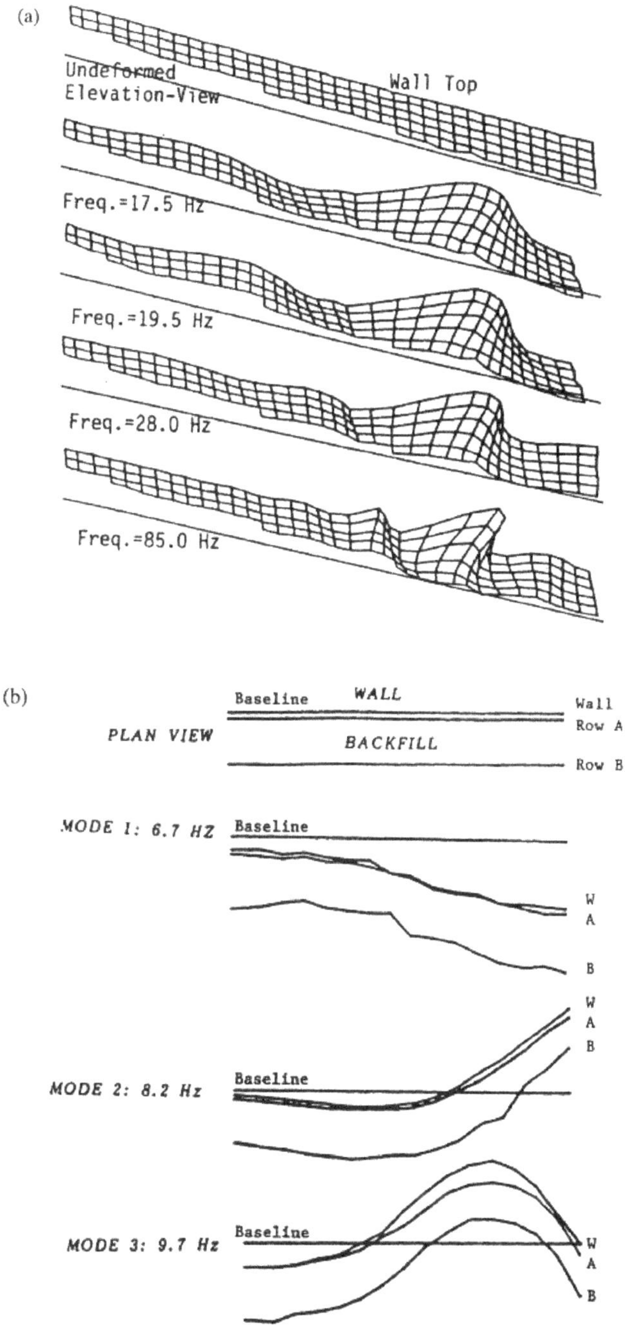

Figure 3: Selected resonant configurations of wall system. (a) In situ impact hammer test and (b) In situ frequency sweep test.

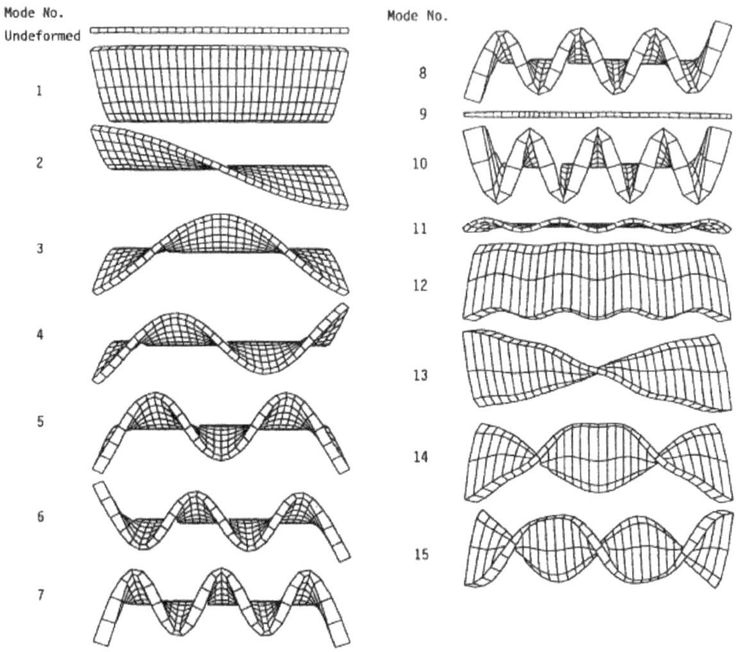

Figure 4: Mode shapes of 30-ft uniform height wall (in plan view).

Uniform Height Wall

The mode shapes of the uniform height wall are shown in Fig. 4, and corresponding modal participation factors are shown in Table 2. (Mode 9 is a longitudinalvertical mode of negligible modal participation factor; hence, attention is focused herein on the other modes.) Only modes 1 and 12 have a significant modal participation factor in the direction perpendicular to the wall face. These modes are the first and second modes along the height, rerspectively, and correspond to modes 1 and 2 of the plane-strain idealization (Table 1).

All other 3-D modes were essentially perfectly symmetric or perfectly antisymmetric along the wall height (Fig. 4). Hence, modal participation factors corresponding to these modes were zero (see Table 2). Such modes would only be excited by nonuniform input ground motion components along the wall length. In general, such motions during an earthquake may be of negligible effect, except for walls of large length.

Table 3: Natural Frequencies and Modal Participation Factors of Wall with Height Varying Linearly from 40 to 20 ft

Mode No.	Frequency (Hz)	Modal Participation Factor*
1	4.459	2.214
2	4.806	−1.172
3	5.271	0.694
4	5.833	−0.376
5	6.752	0.123
6	8.255	0
7	10.391	0
8	12.361	0
9	12.965	−0.807
10	13.173	−0.196
11	15.422	−0.523
12	16.552	0
13	17.594	−0.360
14	19.492	0
15	19.815	0.284

* Motion perpendicular to wall face.

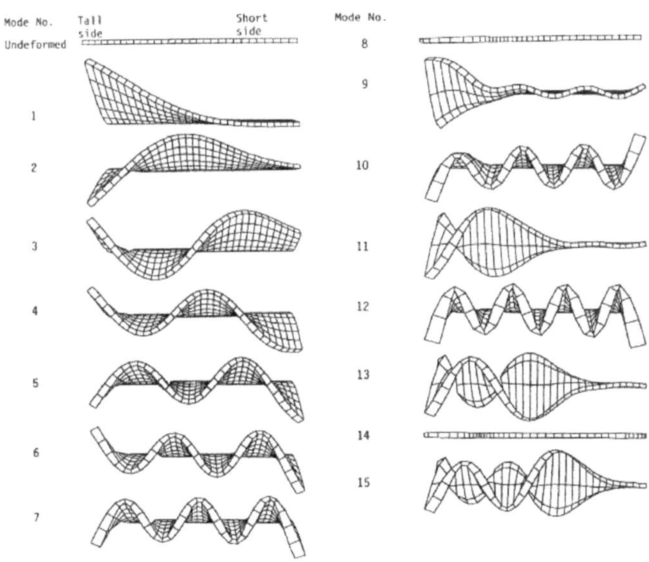

Figure 5: Mode shapes of wall with height varying linearly from 40 to 20 ft (in plan view).

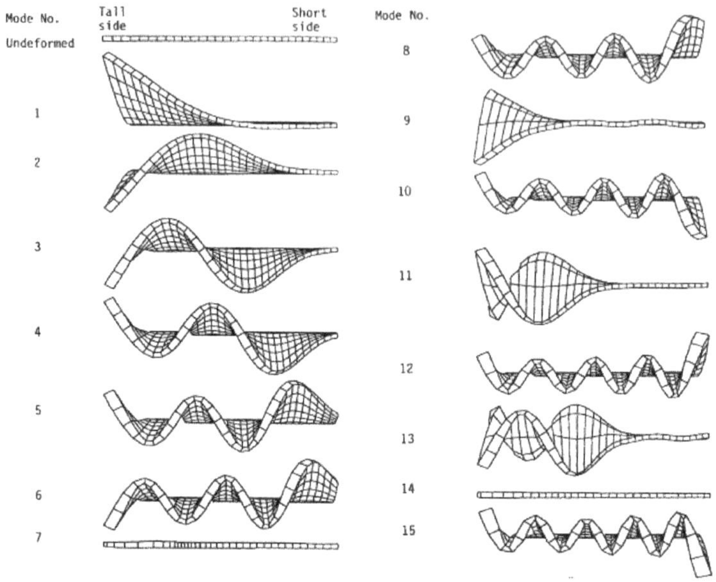

Figure 6: Mode shapes of wall with height varying linearly from 40 to 10 ft (in plan view).

Table 4: Natural Frequencies and Modal Participation Factors of Wall with Height Varying Linearly from 40 to 10 ft

Mode No.	Frequency (Hz)	Modal Participation Factor*
1	4.473	2.185
2	5.038	−1.135
3	5.938	0.687
4	7.211	−0.494
5	8.843	0.381
6	10.765	−0.293
7	12.546	0
8	12.925	0.213
9	13.221	−0.974
10	15.960	−0.139
11	16.680	0.523
12	18.573	0
13	19.988	−0.328
14	21.089	0
15	23.325	0

* Motion perpendicular to wall face.

Nonuniform Height Walls

Mode shapes of these two walls are shown in Figs. 5 and 6, and the corresponding modal participation factors are shown in Tables 3 and 4. (Modes 8 and 14 of Table 3 and Fig. 5

and modes 7 and 14 of Table 4 and Fig. 6 are vertical-longitudinal modes of negligible modal participation factors; hence, attention is focused herein on the other modes.) In the two nonuniform height cases, it is evident that all modes significantly depart from perfect symmetry or asymmetry. This departure increased with the increase in height differential between the left and right ends of the wall. Thus, all modes show significant nonzero modal participation factors. Uniform ground motion will consequently excite all these modes. Under earthquake excitation, any of these modes may playa significant role as dictated by the nature of input ground excitation.

Based on the above discussions, it might be concluded that the 30-ft plane-strain idealization is only effective in matching the first mode of a variable height wall (such as the 40-20 case) and essentially misses all other modes of vibration (Table 3).

Steady-State Forced Response

For above mentioned four wall models, a steady-state forced response analysis was also performed by applying an inertial load in the direction perpendicular to the wall plane. A frequency sweep was conducted in the 0-25 Hz range at an interval of 0.05 Hz (with slight viscous damping to control resonant amplitudes). The frequency sweep (Alampalli, 1990) showed that (Fig. 7) the 30-ft uniform height wall dispayed two main resonances in both its computed 3-D and plane-strain responses. Hence, walls of uniform height can be modeled using plane-strain analysis. In general, contribution of modes involving spatial variation along the length (between the first and second mode given by plane-strain analysis) might be negligible for most typical walls.

Table 5: Natural Frequencies of Uniform Height Wall Using Plate Theory

		Frequency (Hz)		
Mode No.	*m*, *n*	H = 12 ft L = 140 ft T = 1.33 ft	H = 24 ft L = 140 ft T = 2.66 ft	H = 24 ft L = 280 ft L = 2.66 ft
1	1,0	14.23	7.12	7.12
2	1,1	14.62	7.88	7.31
3	1,2	15.80	9.96	7.90
4	1,3	17.55	12.89	8.77
5	1,4	19.75	16.64	9.87
6	1,5	22.43	21.38	11.22
7	1,6	25.60	27.20	12.80
8	1,7	29.28	34.16	14.64
9	1,8	33.48	42.27	16.74
10	1,9	38.23	51.56	19.11

H, plate height; L, plate length; T, plate thickness; plate elastic modulus = 4×10^8 lb/ft^2; Poisson's ratio = 0.15; mass density = 130 lb/ft^3; *m* is the mode number along the height; and *n* is the mode number along the length. Adapted from Wanbourton (1954).

The two walls of variable height (Fig. 7) displayed additional resonances that correspond to modal configurations along the wall length (Figs. 5, 6). These resonances were not excited by the imparted uniform base excitation for the constant height walls (walls I and 4 of Fig. 7). Hence, a dynamic plane-strain analysis of structures such as wing walls (where the height varies continuously along the wall length) may not be satisfactory and a 3-D analysis might be necessary.

As may be noted, only one finite element was used along the wall thickness. The effect of this crude representation along the thickness was investigated (Alampalli, 1990) in four different cases (with 1 to 4 elements along the wall thickness). It was found that number of elements along the thickness had a minor effect on the results reported herein (natural frequencies, mode shapes, and modal participation factors).

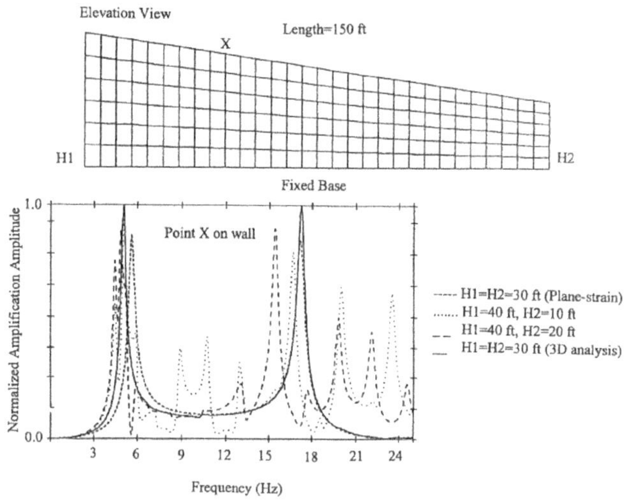

Figure 7: Computed Fourier amplitude spectrum of wall response.

PLATE ANALYSIS

The influence of wall and backfill length and height on dynamic response may be visualized using a simple plate theory idealization (Wanbourton, 1954). In this idealization, the resonances of a wall modeled as a cantilever plate illustrate the effect of length and height (Table 5). The results reveal that the fundamental frequency of the wall is governed by the height and wall length plays no roll in this respect. As the wall length increases, additional low frequency modes appear (along the length). As mentioned earlier, these modes may play an important role in the seismic response of nonuniform height walls.

CONCLUSIONS

The conducted testing program and associated study showed that additional 3-D resonances play an important role in the seismic response of wall backfill soil systems of variable height (such as highway wing walls). For typical uniform height wall systems (with free vertical edges), modal configurations that display spatial variability along the wall length will only be excited by nonuniform seismic motions. Hence, the seismic response of uniform height walls may be adequately analyzed using a 2-D or plane-strain model. Variable height walls are more accurately represented by a 3-D model.

ACKNOWLEDGMENTS

This research was supported by NCEER Contract 90- 1506 under NSF Master Contract ECE-86-07591. Dr. Paul Van Laak played a major role in conducting the full-scale testing program.

REFERENCES

1. Alampalli, S" 1990, "Earthquake Response of Retaining Walls: Full Scale Testing and Computational Modeling," Ph.D. Thesis, Rensselaer Polytechnic Institute, Troy, NY.
2. Alampalli, S., Fu, G., and Dillon, E. w., 1995, "Measuring Bridge Vibration for Detection of Structural Damage," Research Report 165, FHWAlNYIRR-951165, Engineering Research and Development Bureau, New York State Department of Transportation, Albany, NY.
3. Aliev, H., Mamedov, H., and Radgabova, T., 1973, "Investigation of the Seismic Pressure of Soils on the Retaining Walls and Interdependence Between Foundation Soils and Construction," Proceedings of the Symposium on Earth & Earth Structures Under Earthquake & Dynamic Loads, India, pp. 3-10.
4. Amano, R., Azuma, H., and Ishii, Y., 1956, "Aseismic Design of Quay Walls in Japan," Proceedings of the World Conference on Earthquake Engineering, Berkeley, CA, pp.32.1-32.16.
5. Arias, A, Sanchez-Sesma, E J., and Ovando-Shelley, E., 1981, "A Simplified Elastic Model for Seismic Analysis of Earth-Retaining Structures with Limited Displacements," Proceedings of the International Conference on Recent Advances in Geotechnical Earthquake Engineering and Soil Dynamics, Rolla, MI, pp. 235-240.
6. Calciati, E, Castoldi, A., Fanelli, M., and Mazzieri, C., 1977, "In Situ Tests for Determination of the Dynamic Characteristics of Some Italian Dams," Proceedings of the Sixth World Conference on Earthquake Engineering, Vol. III, India, pp. 2823-2828.
7. Chang, C. Y., Power, M. S., Mok, C. M., Tang, Y. K., and Tang, H. T., 1990, "Analysis of Dynamic Lateral Earth Pressures Recorded on Lotung Reactor Containment Model Structure," Proceedings of the 4th National Conference on Earthquake Engineering, Vol. 3, Palm Springs, CA, pp. 643-652.
8. Douglas, B. M., and Reid, W. H., 1982, "Dynamic Tests and System Identification of Bridges," Journal of Structural Division, ASCE, Vol. 108, No. STlO, pp. 2295-2312.

9. Elgamal, A-W., Alampalli, S., and Van Laak, P., 1996, "Forced Vibration of Full-Scale Wall-Backfill System," Journal of Geotechnical Engineering Division, ASCE, Vol. 122, No. 10, pp. 849-858

10. Elgamal, A-W., Van Laak, P., and Alampalli, S., 1990, "Modal Response of Earth Retaining Structures," Proceedings of the Eighth International Modal Analysis Conference, Kissimmee, FL, Vol. I.

11. Ewins, D. 1., 1984, Modal Testing: Theory and Practice, Wiley, New York.

12. Fukuoka, M., and Imamura, Y., 1984, "Researches on Retaining Walls During Earthquakes," Proceedings of the Eighth World Conference on Earthquake Engineering, San Francisco, CA, Vol. 3, pp. 501-508.

13. Jennings, P. C., 1971, Engineering Features of San Fernando Earthquake, February 9, 1971, Earthquake Engineering Research Laboratory, California Institute of Technology, Pasadena, CA

14. Keightley, W. 0., 1966, "Vibrational Characteristic of an Earthdam," Bulletin of Seismological Society of America, Vol. 56, No.6 pp. 1207-1226.

15. Matsuo, H., and Ohara, S., 1960, "Lateral Earth Pressure and Stability of Quay Walls During Earthquakes," Proceedings of the Second World Conference on Earthquake Engineering, Tojyo, Japan, Vol. 1, pp. 165-181.

16. Paska10v, T. A, Petrovski, J. T., Jurukovski, D. v., and Taskovski, B. M., 1977, "Forced Vibration Full-Scale Tests on Earth-Fill and Rock-Fill Dams," Proceedings of the Sixth World Conference on Earthquake Engineering, India, Vol. III, pp. 2829-2834

17. Seed, H. B., and Whitman, R. v., 1970, "Design of Retaining Structures for Dynamic Loads," Lateral Stresses in Ground and Design of Earth Retaining Structures, Specialty Conference, pp. 103-145.

18. Sitar, N., 1995, Geotechnical Reconnaissance of the Effects of the January 17,1995, Hyoken-Nanbu Earthquake, Japan, Report UCBIEERC-951 10 1 , Univ. of California at Berkeley, Berkeley, CA.

19. SMS, 1987, Moda13.0SE Modal Analysis System, Structural Measurement System, Inc., CA.

20. Tatsuoka, E et al., 1996, "Performance of Soil Retaining Walls for Railway Embankments," Soils and Foundations, Special Issue on Geotechnical Aspects of the January 17, 1995, Hyogo-ken Nanbu Earthquake, Japan, Japanese Geotechnical Society, No. 1.

21. Velestos, A S., Parikh, V. H., and Younan, A H., 1995, "Dynamic Response of a Pair of Walls Retaining a Viscoelastic Solid," Earthquake Engineering and Structural Dynamics, Vol. 24, pp. 1567-1589

22. Wanbourton, G. B., 1954, ''The Vibration of Rectangular Plates," Proceedings of the Institution of Mechanical Engineers, Vol. 168

23. Whitman, R. V., 1991, "Seismic Design of Earth Retaining Structures (State of the Art Paper)," Proceedings of the Second Inti. Conference on Resent Advances

in Geotechnical Earthquake Engineering and Soil Dynamics, in S. Prakash, St. Louis, MO.

24. Whitman, R. v., and Christian, J. T., 1990, “Seismic Response of Retaining Structures,” Symposium on Seismic Designfor World Port 2020, Port of Los Angels, San Pedro, CA

25. Wood, J. H., 1973, “Earthquake Induced Soil Pressures on Structures” , Ph.D. Thesis, California Institute of Technology, Pasadena, CA

26. Wood, J. H., 1975, “Earthquake Induced Soil Pressures on Structures”, Bulletin of New Zealand Society of Earthquake Engineering, Vol. 8, pp. 175-186.

11

CHAPTER

THE EFFECT OF DISPLACEMENT MODE OF RIGID RETAINING WALLS ON SHEARING BANDS BY ACTIVE EARTH PRESSURE

A. Sekkel, M. Meghachou

Keywords: Small-scale model, active earth pressure, continuous mode, discontinuous mode, shearing band

INTRODUCTION

Earth pressure problems are always encountered in retaining structures such as retaining walls, sheet pile bulkheads, cofferdams, bridge abutments, basement walls of buildings etc. Evaluation of earth pressure is of practical significance for the design of these retaining structures and it is a key research subject in soil mechanics and geotechnical engineering. Although results of full-scale model tests are more trustworthy, the costs to conduct such tests are very high. Scale model is a powerful tool for the verification of the earth pressure theory and the investigation of the mechanism of earth pressure. A lot of research on small-scale model was carried out by Terzaghi into 1934 which is a big step in the history of the earth pressure study [1]. The results obtained showed that wall displacement has a significant effect on earth pressure. The earth pressure distribution on rigid retaining walls rotating about the base through field full-scale tests has been investigated [2]. A model of retaining wall has been developed in order to study the effect of wall displacement mode on earth pressure distribution [3]. The effect of the angle of inclination of the wall on the pressure of the grounds has also been studied [4].

Despite intense theoretical and experimental researches for more than 200 years, there remain considerable discrepancies between theoretical solutions and experimental results, due to the complexity of deformation field in granular bodies near the wall caused by the localization of the deformations (which is a fundamental phenomenon of

Citation: Abdelkrim Sekkel, Mourad Meghachou, The Effect of Displacement Mode of Rigid Retaining Walls on Shearing Bands by Active Earth Pressure, ISSN: 2241-4487 (Print); 1792-8036.

granular material behavior). The localizations can appear single or multiple, depending upon initial and boundary conditions. They can be plane or curved. In shearing zones, pronounced grains rotations [5-8] and large strain gradients [9] are observed. The thickness of shear localizations depends on various factors such as: mean grain diameter [9, 7, 10-12], pressure level [10, 13], initial void ratio [7, 13], direction of deformation [11], grains roughness and grain size distribution [7, 10, 11].

Coulomb [14] indicated for the first time, the occurrence of shear zone during earth pressure tests. Darwin [15] illustrated that the explanation of the behavior of the granular bodies during earth pressure is not possible without taking into account shear localization. Several comprehensive experimental studies on earth pressure in sand were carried out between 1962 and 1974. During this period, a number of researchers [16-24] carried out experiments on the active and passive failure of dry sand deforming under plane strain conditions. The type of wall movement (passive wall translation) [18], active wall rotation about its top [22], passive wall rotation about its top [16, 17, 22], active wall rotation about its bottom [23, 24], passive wall rotation about its bottom [20, 21], wall roughness [17, 24], wall flexibility (rigid wall and flexible wall) [24], initial density of sand (dense and loose) [17, 22], and influences of surcharge, were studied. The evolution of shear localization in sand was recognized by using the radiographic technique which is able to detect density changes. Different modes of shear zone during passive and active earth pressure tests depend mainly on the type of wall motion and surcharge. In passive tests with rigid walls rotating about the top, one or two curved shear zones were obtained in sand. Multiple curved shear zones of a similar shape were observed during tests with a wall rotating about the bottom. They occurred at the wall top and propagated towards the free boundary. During tests with translating rigid walls, slightly curved shear zones starting to form from the wall bottom, and secondary radial shear zones appears at the wall top. In active tests with rigid walls, nearly parallel straight zones or a mesh of intersecting parallel zones close to the wall (wall rotating around the bottom) or a single curved zone (wall rotating around the top) were observed. The details of these along with radiographs can be found in [25].

Numerical calculations [26, 27] revealed that the pattern of shear zones were easier to observe when a non-associated model of Coulomb-Mohr without dilatancy was assumed. The study carried out by Tejchman [28] with a uniform distribution of the initial index of the initial void ratio in sand showed that the calculated deformation field with a wall rotating around the bottom (passive and active case) was different from the experimental one [20, 23]. Further, this study showed that the geometry of calculated shear zones was influenced by the size of the sand specimen and its initial void ratio. The pattern of the induced shear zones was created also in the case of loose sand continuously subject to contractancy. This paper wants to answer the following question: How shear zones will evolve when modifying the wall displacement mode? Will they follow logic evolution or random diffusion? Wall movement is translating in the X-X plane in continuous and discontinuous mode. So that the problem is mathematically determined, it is necessary to also set the boundary conditions on the two borders, free face and screen. The study will be related to horizontal free face not surcharged and vertical smooth wall, and this, for all tests. The studied material is that of Schneebeli; light twodimensional material made up of dense rearrangement of cylindrical plastic rollers, simulating granular soil non-cohesive (dense sand).

EXPERIMENTAL PROGRAM

Description of active earth pressure was concretized by suitable experimental study. If one of assumptions theories of states ultimate equilibrium is that of a two-dimensional massif, this assumption cannot be ignored during any experimental work. The use of correlation pictures (Kisdef-7D software), allows qualitative visualization and especially the quantification of fields displacement evolution and those of plane objects deformations which will remain plane after transformation process. The principle of method consists in comparing two digital images. A first recording of granular soil surface is taken before deformation, whereas a second is taken after deformation [29]. The design of a small-scale model could as well as possible correspond to physical reality, which remains nevertheless very complex. It is thus a question of studying a physical phenomenon on a reduced dimensional reproduction in smallscale model [30].

Modeling of Granular Soil

The studied granular soil is represented by Schneebeli material which is used in various experimental approaches [5, 31]. Schneebeli in 1957 showed that problems in soil mechanics with a granular non-cohesive material can be reduced to two dimensions [32]. The analogy consists in replacing a non-cohesive soil by a stacking of cylindrical, twodimensional rollers.

A lot of biaxial tests were carried out with cylindrical rollers made by dural [32-35] (dural or duraluminium is an alloy of aluminum and copper used in aeronautics because of its great strength and its more weak density that steel), rubber [36], PVC [37-40] and steel [41]. All tests showed that the material behavior is identical to dense sand. The material studied in this paper, consisted of light plastic cylindrical rollers accumulation (Eighteen thousand rollers), as shown in Figure 1, cut out carefully showing the following characteristics:

- Light massif, dry density, γ_d= 3,5 kN/m3 , grain density γ_s= 9 kN/m^3 with plasticine, its cohesion is null and its natural friction angle ϕ=10°.
- Indeformable rollers within sight of the actions applied thus simulating incompressible grains. The initially cylindrical rollers remain cylindrical after a process of transformation;
- Those rearrangements confer on granular soil a dense configuration.
- Length and diameter are respectively (70x2) in mm.

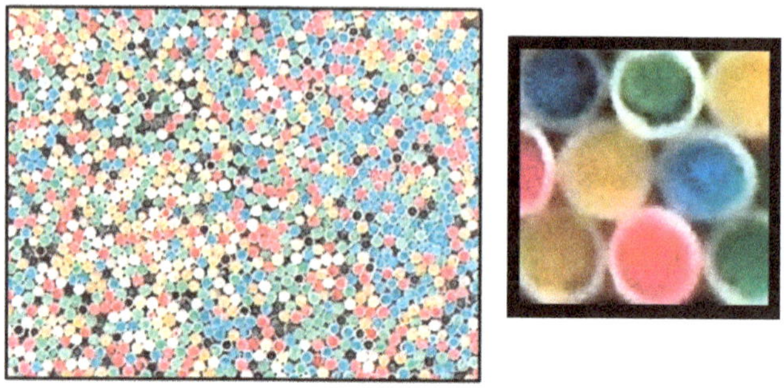

Figure 1: Dense configuration of plastic cylindrical rollers.

Rollers have an intrinsic porosity which is not held in account in our estimate. This is justified by the assumption of rollers incompressibility compared to implemented solicitations.

Design of a Small-scale Model

Figure 2 illustrates experimental small-scale model. Dimensions for the length, width and depth are respectively 800 mm, 500 mm and 100 mm. Thus, a displacement wall of 30 mm is relatively low compared to the length justify a semiinfinite massif. Shear bands, when propagated, are not blocked by any other side wall. The 100 mm depth is sufficient to allow filling under the best stability conditions with rollers witch occupied a volume of (550x250x70) mm^3 .

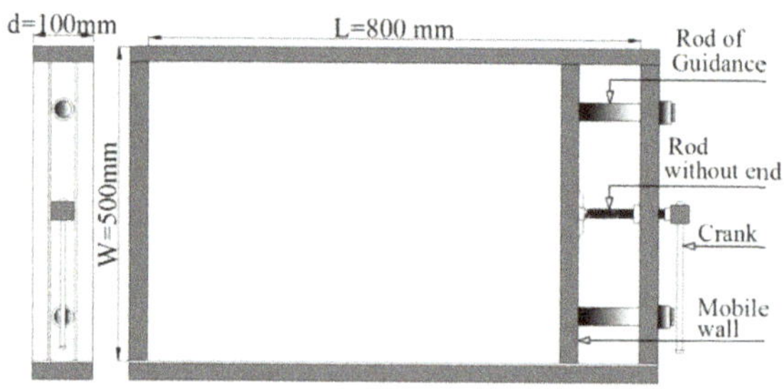

Figure 2: Experimental small-scale model (Ref : 01-DPB-040404).

Tests Process

Tests are carried in aim of qualitative comparison of rupture evolution between two displacement modes of rigid retaining wall known as continuous and discontinuous. Continuous mode describes wall displacement in an uninterrupted manner during all test process whereas discontinuous mode describes wall displacement marked by regular stops.

Tests are structured as follows:

- Wall Displacement in continuous mode: mobile wall carried out a translation in the X-X plane at constant speed of 0,22 mm/s without interruption in time during tests process. Photographs are taken in a continuous way during all transformation process. Total displacement of mobile wall is 30 mm.
- Wall Displacement in discontinuous mode: analog with continuous mode, however the mobile wall marks an interval of stop and a downtime; interval of stop is estimated at 0,75 mm when at the downtime, it is 3 seconds, and then the wall starts again.
- After each test, rollers are rearranged and compacted to eliminate dilatancy zones to have the same initial density. The depth must be maintained within 250 mm. For each mode of wall displacement, the tests are repeated five (5) times by taking photographs with steps identical to tests preceding them.
- A post treatment by the software of correlation pictures Kisdef-7D allows the quantification and shearing bands evolution for studied modes.

Result Analysis by Correlation Pictures

The treatment of active earth pressure tests is done by Kisdef-7D and it focuses in small deformations analysis by correlation pictures process. Digital pictures materializing transformations states are introduced into the software's database. Domain analysis is defined by the operator; it is subdivided in quadrilaterals by a virtual grid of 10x10 pixels on the initial image (see Figure 3). Deformed grid is found by comparing gray levels for each couple of pictures treated with high accuracy (10^{-8} pixels or $0,19x10^{-8}$ mm).

It should be noted that picture resolution (see Figure 4) is 3264x1836 pixels. Its confer ratio mm/pixel of 0,19.3264 pixels correspond at 620 mm in actual size. 620 mm/3264 pixels=0,19 thus 1pixel=0,19mm.

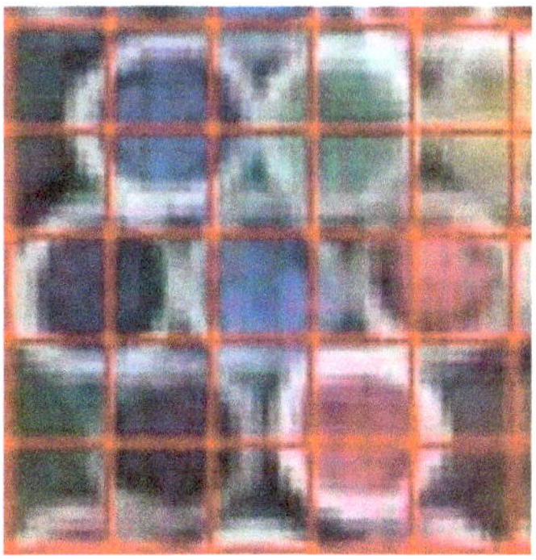

Figure 3: Definition of quadrilaterals (10x10) pixels.

Figure 4. Ratio mm/pixel.

Analysis Mode and Scale Effect Representation

Shearing band is evolutionary for increment displacement and its research requires choosing an incremental type analysis. Therefore, photograph couples will be analyzed in incremental way i.e. (1-2), (2-3), (3-4), (49-50)… etc (see Figure 5). Regarding scale effect representation, the software has two alternatives: local scale and total scale. Local scale proposes min and max values of maximum shearing for each analyzed couple, therefore varies from couple to couple. Total scale, proposes min and max values of maximum shearing for all analyzed couples.

All tests in continuous and discontinuous mode will show fluctuations in shearing bands intensities during transformation process, which will be visible only in local scale, as shown in Figure 5. Therefore, local scale will be adopted for each treated couple, as a total scale on all pictures couples will not allow observation of shear bands localization. On analysis 49 treating photographs 49 and 50 in continuous mode, shearing band evolution is definitely clearer on left figure (local scale), that on right figure (total scale). The same parameters are valid for discontinuous mode

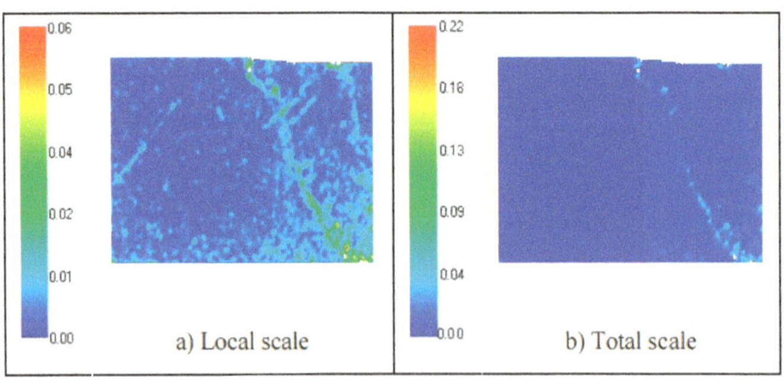

Figure 5: Scale effect representation on localization shearing band: analyze 49/49 between photographs 49 and 50: Maximum shearing.

Evolution of Shearing Bands in Continuous and Discontinuous mode

A question however arises: "How shearing bands will evolve by modifying wall displacement mode? Will they follow logic evolution or random diffusion?" Figure 6 represents failure evolution of granular soil following active earth pressure by shearing bands. A comparison between the two modes makes it possible to observe rupture evolution during transformation process.

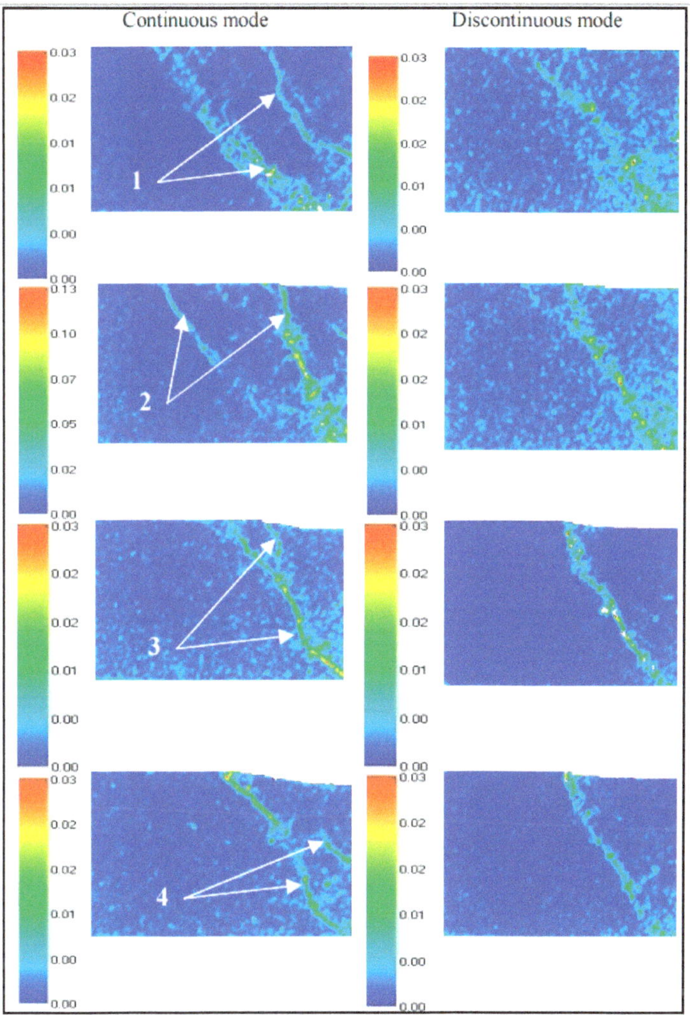

Figure 6: Shearing bands evolution in continuous and discontinuous mode.

When a granular soil is solicited in active earth pressure, a localization of deformation appears by shearing bands. Dilatancy phenomenon for dense soils is inevitable. The relative grain slips modify their rearrangement and the triangular configuration (dense state) become square (loose state) localized in the shearing band.

Density distribution is not homogeneous (low in shearing band). When we continue to solicit soil in active earth pressure, it is similar to starting from an initial configuration with a heterogeneous distribution density, therefore slip is favored in the low density zone. If the process is reiterated, the deformation localization will settle more and more in the area in question.

However, for continuous mode displacement, in active earth pressure equivalent to a discontinuous sum of displacement mode, shearing bands propagation follows a diffusion logic letting multi mechanisms of failure appear (see Figure 6, labels 1-4).

Shearing Bands Evolution Before and after Each Stop of wall Movement

Figure 7 illustrates shearing bands evolution for discontinuous mode displacement before and after each stop of mobile walls. Results are expressed in maximum shearing.

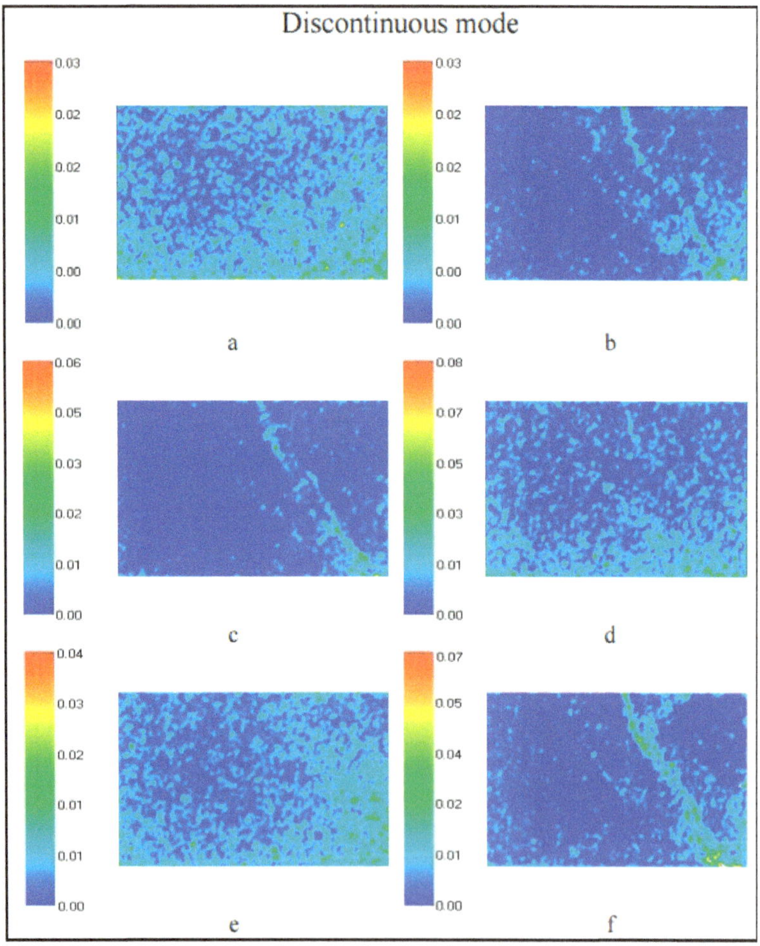

Figure 7: Evolution of shearing bands before and after stop of wall movement.

As shown in Figure 7a, at stop of wall movement, an equilibrium state is created in granular soil. With movement wall resuming, shearing band appears clearly (Figure 7b) and continues (Figure 7c). At time of second stop, a new equilibrium state is created with an heterogeneous density configuration (Figures 7d-7e). The last figure is spectacular; its shows the reappearance of shearing band exactly at the same place where it had disappeared when preceding wall movement had stopped.

Repeatability

For results validation, six repeatability tests are carried out under same conditions and lead to the same conclusions, as shown in Figure 8.

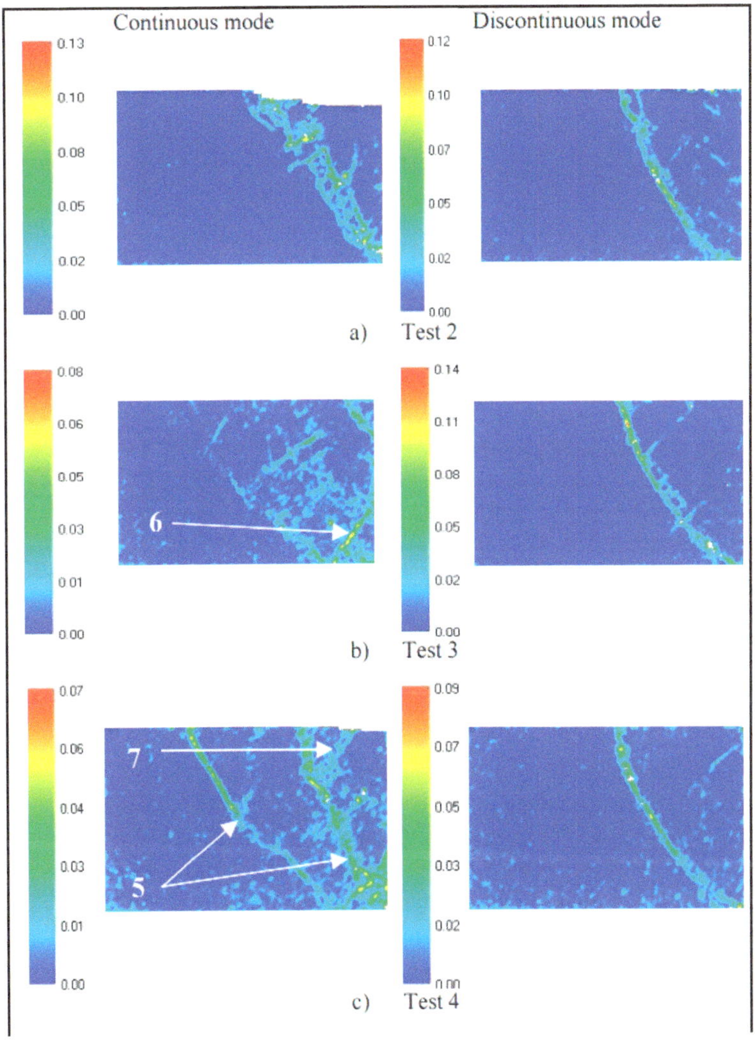

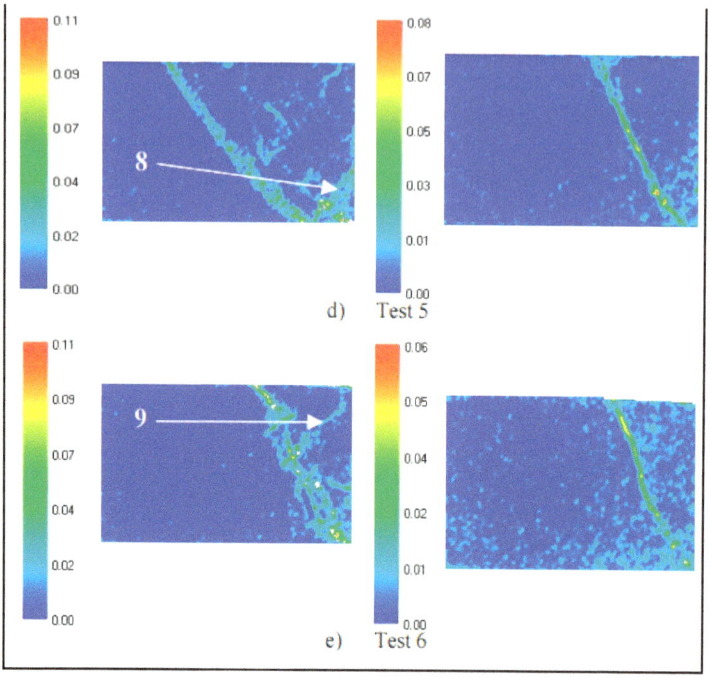

Figure 8: Tests validation.

Tests results expressed in maximum shearing intensity during discontinuous mode, (on the right of Figure 8) clearly show the preponderance of shearing band, while results on the left for continuous mode represent a disturbance of massif, governed by multiple shearing bands accompanied by parallel bands (Figure 8c, label 5), as well as perpendicular bands (Figure 8b-8c-8d-8e, labels 6-9). The difference is clearly shown.

CONCLUSION

At the last of this work, we arrive at the following conclusions:

- Realization of small-scale model simulating a granular soil subjected to active earth pressure made it possible to often highlight the evolution of deformations localization by shearing bands
- During the tests known as discontinuous, shearing bands evolution in dense massif favored dilatancy phenomenon, governs by low density in sheared zone. With the resumption of wall movement, the starter of a new shearing band is favored in sheared zone which is the least dense of the massif. This results strong deformation localization.
- With each stop of wall movement, an equilibrium state is created with a heterogeneous density configuration. With each resuming of wall movement, granular soil is solicited again and shearing band reappears exactly at the same place where it had disappeared at time of stop preceding wall movement.

- During tests known as continuous, shearing bands evolution follows logic diffusion letting appear strong disturbance governed by several shearing bands translating multi-mechanisms of failure.
- It was shown through this paper that failure evolution is significantly affected by displacement mode of mobile wall. The qualitative study shows a clear difference between continuous and discontinuous mode. This difference can be thorough in prospect while varying the interval of stop, the downtime as well as displacement speed of rigid retaining wall.

REFERENCES

1. K., Terzaghi, "Large retaining wall tests", Engineering News Record Vol. 112, pp. 136-140, 1934
2. M. Matsuo, S. Kenmochi, H. Yagi, "Experimental study on earth pressure of retaining wall by field tests", Soils and Foundations, Vol. 18, No. 3, pp. 27–41, 1978
3. M. A. Sherif, Y. S. Fang, R. I. Sherif, "Ka and K0 behind rotating and non-yielding walls", Journal of Geotechnical Engineering Vol. 110, No. 1, pp. 41-56 , 1984
4. Y. S. Fang , T. J. Chen, B. F. Wu, Passive earth pressures with various wall movements, Journal of Geotechnical Engineering, Vol. 120, No. 8, pp. 1307-1323, 1994
5. M. Oda, J. Konishi, S. Nemat-Nasser, "Experimental micromechanical evaluation of strength of granular materials: effects of particle rolling", Mechanics of Materials, Vol. 1, pp. 269-283, 1982
6. M. Uesugi, H. Kishida, Y. Tsubakihara, "Behaviour of Sand Particles in Sand-Steel Friction", Soils and Foundations, Vol. 28, No. 1, pp. 107– 118, 1988
7. J. Tejchman, "Scherzonenbildung und Verspannungseffekte in Granulaten unter Berucksichtigung von Korndrehungen", Publication Series of the Institute of Soil and Rock Mechanics, University Karlsruhe, Vol. 117, pp. 1–236, 1989
8. J. Geng, R. R. Hartley, D. Howell, R. P. Behringer, G. Reydellet, E. Clement, "Fluctuations and instabilities in granular materials" in: Bifurcations and Instabilities in Geomechanics, (J. Labuz and A. Drescher, eds.), Swets and Zeitlinger, 79–108, 2003
9. I. Vardoulakis, Shear band inclination and shear modulus of sand in biaxial tests", International Journal for Numerical and Analytical Methods in Geomechanics, Vol. 4, No. 2, pp. 103–119, 1980
10. F. Tatsuoka, M. Okahara, T. Tanaka, K. Tani, T. Morimoto, M. S. Siddiquee, "Progressive failure and particle size effect in bearing capacity of footing on sand", Proc. of the ASCE Geotechnical Engineering Congress, Vol. 27, No. 2, pp. 788–802, 1991
11. F. Tatsuoka, M. S. Siddiquee, T. Yoshida, C. S. Park, Y. Kamegai, S. Goto, Y. Kohata, "Testing methods and results of element tests and testing conditions of

plane strain model bearing capacity tests using airdried dense Silver Buzzard Sand", Internal Report of the University of Tokyo, 1–129, 1994

12. F. Tatsuoka, S. Goto, T. Tanaka, K. Tani, Y. Kimura, "Particle size effects on bearing capacity of footing on granular material", in: Deformation and Progressive Failure in Geo-mechanics, (A. Asaoka, T. Adachi, and F. Oka, eds.), Pergamon, pp. 133–139, 1997
13. J. Desrues, W. Hammad, "Shear banding dependency on mean pressure level in sand", Proc. Int. Workshop on Numerical Methods for Localization and Bifurcation of Granular Bodies, (E. Dembicki et al, eds.), Gdańsk, Poland, pp. 57–67, 1989
14. C. A. Coulomb, Essai sur une application, Science et Industrie, Paris, 1775
15. G. H. Darwin, "On the horizontal thrust of a mass of sand", Proc. Inst. Civ. Eng., Vol. 71, pp. 350–378, 1883
16. J. R. F. Arthur, Strains and lateral force in sand, PhD Thesis, University of Cambridge, 1962
17. R. G. James, Stress and strain fields in sand, PhD Thesis, University of Cambridge, 1965
18. J. B. A. Lucia, Passive earth pressure and failure in sand, Research Report, University of Cambridge, 1966
19. J. May, A Pilot Project on the cutting of soils, Research Report, University of Cambridge, 1967
20. P. L. Bransby, Stress and strain in sand caused by rotation of a model wall, PhD Thesis,University of Cambridge, 1968
21. A. Adeosun, Lateral forces and failure patterns in the cutting of sands, Research Project, University of Cambridge, 1968
22. J. A. Lord, Stress and strains in an earth pressure problem, PhD Thesis, University of Cambridge, 1969
23. I. Smith, Stress and strain in a sand mass adjacent to a model wall, PhD Thesis, University of Cambridge, 1972
24. G. W. E. Milligan, The behaviour of rigid and flexible retaining walls in sand, PhD Thesis, University of Cambridge, 1974
25. D. Leśniewska, Analysis of shear band pattern formation in soil. Habilitation, Institute of Hydro-Engineering of the Polish Academy of Sciences, Gdańsk, Poland, 2000
26. D. Leśniewska, Z. Mróz, "Study of evolution of shear band systems in sand retained by flexible wall", International Journal for Numerical and Analytical Methods in Geomechanics, Vol. 25, No. 9, pp. 909–932, 2001
27. D. Leśniewska, Z. Mróz, "Shear bands in soil deformation processes", in: Bifurcations and Instabilities in Geomechanics (J. Labuz and A. Drescher, eds.), Swets and Zeitlinger, pp. 109–119, 2003

28. J. Tejchman, "Patterns of shear zones in granular materials within a polar hypoplastic continuum", Acta Mechanica, Vol. 155, No. 1–2, pp. 71–95, 2002

29. P. Vacher, S. Dumoulin, F. Morestin, S. Mguil-Touchal, "Bidimensional strain measurement using digital images", Journal of Mechanical Engineering Science, Vol. 213, No. 8, pp. 811-817, 1999

30. F. Darve, "Manuel de rhéologie des géomatériaux", Edition de l'association Amicale des Ingénieurs Anciens Elèves de l'Ecole Nationale des Ponts et Chaussées, 1987

31. J. Lanier, P. C. Bonnet, "Experimental study of displacements and rotations of grains in 2D-granular media", Meeting TC13, Comitee on mechanics of granular materials (ISSMFE), Birmingham, 1993

32. G. Schneebeli, "Une mécanique pour les terres sans cohésion", Compte rendus des séances de l'Académie des Sciences. Paris, Vol. 243, pp. 2647-2673, 1956

33. C. Coulet, "Etude du profil optimal des murs-voiles de soutenement", Thèse, Lyon : INSA de Lyon, 1977

34. R. Albert, A. Houy, J. Raud, G. Verdier, "Etude à l'aide de modèles réduits de Datardeaux Massifs", Proc. of 5 Congrès Européen, Madrid, Vol. 1, pp. 231-239, 1972

35. J. Verdeyen, V. Roisin, "Sollicitation des écrans souples de soutènement", in : Proc. Of 5 Congrès Int. , Paris, Vol. 1, pp. 422-426, 1961

36. M. Oda, J. Konichi, S. Nemat-Nasser, "Experimental micromechanical evaluation of the strength of granular materials: effects of particle rolling", in. : Jenkins and Satake Eds. Mechanics of Granular Materials: New Models and Constitutive Relations, pp. 21-30, 1983

37. R. Gourves, M. Mezghani "Micromécanique des milieux granulaires, approche expérimentale utilisant le modèle de Schneebeli", Revue Française de Géotechnique, Vol. 42, pp. 23-34, 1988

38. H. Joer, J. Lanier, J. Desrues, E. Flavigny, "1γ2□: A new shear apparatus to study the behavior of granular materials", Geotechnical Testing Journal, Vol. 15, No 2, pp. 129-137, 1992

39. G. Tratapel, "Contribution à l'étude du comportement mécanique d'un milieu analogique bidimensionnel", Thèse: Génie Civil : Université Scientifique et Médicale et l'Institut National Polytechnique de Grenoble, 1977

40. A. Misra, H. Jiang, "Measured kinematic fields in the biaxial shear of granular materials", Computers and Geotechnics, Vol. 20, No. 3-4, pp. 267-285, 1997

41. R. Kastner, "Excavations profondes en site urbain", Thèse Génie Civil, Lyon : INSA de Lyon, 1982

12
CHAPTER

NUMERICAL SIMULATION OF TIRE REINFORCED SAND BEHIND RETAINING WALL UNDER EARTHQUAKE EXCITATION

Amina Lazizi, Habib Trouzine, Aissa Asroun, Farid Belabdelouahab

Keywords: Waste tire, sand, retaining wall, earthquake, simulation

INTRODUCTION

With the rapid growth in automobile usage worldwide, approximately 2–3 billion tires are discarded annually [1-2]. Algeria alone produces approximately 25918.50 tons/year of used tires [3]. The most common ways to recycle tires are burying, auxiliary fuel conversion, pyrolysis, reproduction, cutting and reprocessing, and reusing. Applications in geotechnical engineering, highway engineering, soil and water conservation, and marine engineering use intact used tires. Such applications which do not require complex processes are considered economical and eco-friendly [4-6].

The first research in France on soil reinforcements using old tires was commissioned in 1976. Studies were conducted on reinforcement in the form of whole tires, sidewalls, or treads placed on edge or cut and laid flat [4-7]. The first project in England using scrap tires was the construction of an experimental gravity wall in West Yorkshire [8]. An experimental research using a vibrating table was also undertaken from the "l'Ecole Nationale des Ingénieurs deTunis" (ENIT) and the "Laboratoire Centrale des Ponts et Chaussés" (LCPC), whose objective was to protect the Grenoble synchrotron road from vibrations. This study showed that used tires can attenuate vibration phenomena [9]. Another full-scale experimentation was conducted in Rouen in France; this research showed that tire reinforced soil can absorb and reduce pressure and acceleration [10].

In the seismic stability analysis, the magnitude of the dynamic force increment due to shaking is calculated using the Mononobe-Okabe (M.O) approach [11]. Basha

Citation: Amina Lazizi, Habib Trouzine, Aissa Asroun, Farid Belabdelouahab, Numerical Simulation of Tire Reinforced Sand Behind Retaining Wall Under Earthquake Excitation, ISSN: 2241-4487.

and Babu in [12] presented a method to evaluate the internal stability of reinforced soil structures against tension and pullout modes of failures using pseudo-static methods for earthquake conditions. Further, Choudhury and Ahmed in [13] presented the external stability of waterfront reinforced soil structures under seismic conditions using a pseudo-static method and a limited equilibrium approach. Steedman and Zeng in [14] developed the pseudo-dynamic method considering horizontal seismic acceleration and finite shear waves propagations using planar failure surface.

In the present paper, a numerical simulation of tire reinforced sand behind retaining wall under earthquake using implicit and explicit methods is performed in order to study its dynamic behavior.

DESCRIPTION OF THE NUMERICAL MODEL

Geometry and Ground Conditions

Geometrically, the retaining wall is a reinforced concrete cantilever wall with 5.80 m of height and 0.50 m of wall stem thickness (Figure 1). Mechanically, the Young's modulus of the retaining wall is E=3.107 kN/m^2 , the Poisson's ratio is about 0.3 and its density is about 26.43 kN/m^3 . Table I summarizes all the characteristics. The earthfill is a sandy soil with a semi-infinite length of 22.50 m. The surface of the backfill is plane and carries no surcharge. The seismic action is introduced in the form of the accelerogram acquired from the earthquake dated May 18, 1940 in El Centro in the region of California.

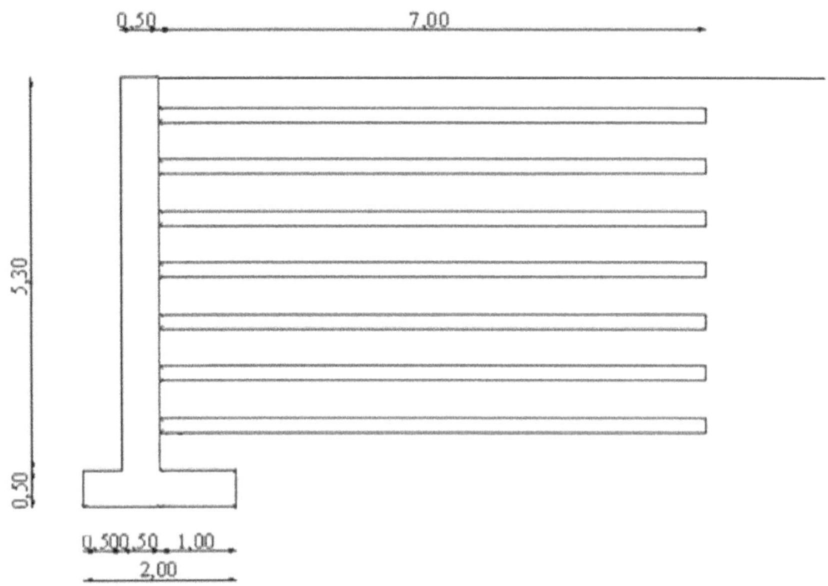

Figure 1: Geometric details of the wall.

Table I: Model Characteristics

Materials	Properties
Retaining Wall	E_w = 3.E+7 kN/m² υ_w=0.25 ρ_w=26 430 kg/m³ damping α =0.015
Sand	E_s = 100 000 kN/m² υ_s=0.25 ρ_s=18 260 kg/m³ damping α =0.01
Tire reinforced Sand	E_t = 100 000 kN/m² υ_t=0.25 ρ_t=17 940 kg/m³ damping α =0.02

Studied Cases

Two distinct cases are studied; a reference case where the wall supports only the sand and a second case where the sand is reinforced by layers of waste tires. Both cases are studied under static and dynamic loading.

Case 1: Retaining wall Supporting Sand

The backfill is wet sand with an average diameter of 1.10 mm, a density of 26.5 kN/m^3 , and a coefficient of curvature about 0.119 to 0.136. According to the unified classification (USCS), the soil is SP poorly graded clean sand. Geotechnically, the cohesion is zero, and the angle of internal friction is 30°. Mechanically, the Young's modulus E is 1.105 kN/m^2 , the Poisson's ratio is about 0.25 and its density is 18.40. The active earth pressure coefficient is determined using the Rankine method, and it is about 0.333. The wall slip is considered with a coefficient of friction of 0.4.

Case 2: Retaining Wall Supporting Tire Reinforced Sand

In this case, the same sand is reinforced with layers of light vehicle tires type flanks without treads on edge. The composite (tire-sand) is amended as multi layer, its Young modulus and Poisson's ratio are determined using Voigt Reuss combined model:

$$E_{C-VR} = E_m.\beta / \gamma \qquad (1)$$

$$\beta = E_m + V_a^{2/3}\left(E_a - E_m\right) \tag{2}$$

$$\gamma = \beta\left(1 - V_a^{1/3}\right) + \left(E_m \cdot V_a^{1/3}\right) \tag{3}$$

with:

$E_{C\,VR}$ - : the Equivalent Young's modulus

E_m, E_a : the Sand and Waste Tire Young moduli

$V_a = \dfrac{V_c}{V_c + V_s}$ with , V_c V_s : the sizes of tire and sand

The tire height is 20 cm, and the layer of sand is 50 cm. The reinforcements are linear modeled as a superposition of the layer of sand and waste tire. The sand-rubber interaction is considered with a coefficient of sliding friction of 0.6; the rubber-wall friction coefficient is equal to 0.65. The active earth pressure coefficient is computed the same way as in the first case.

STATIC AND SEISMIC LOADING

For static loading, the weight of retaining wall is computed, and the active pressure is calculated using the Rankine method and is about 86 kN in the first case and 85.4 kN in the second case. For seismic loading, the accelerogram of the El Centro Earthquake in California is introduced in the form of a table of two columns, the first for time, which is limited in 14 s, and the second for the values of acceleration [15]. Rayleigh mass proportional damping factor is introduced for each element (wall, soil and tire-soil).

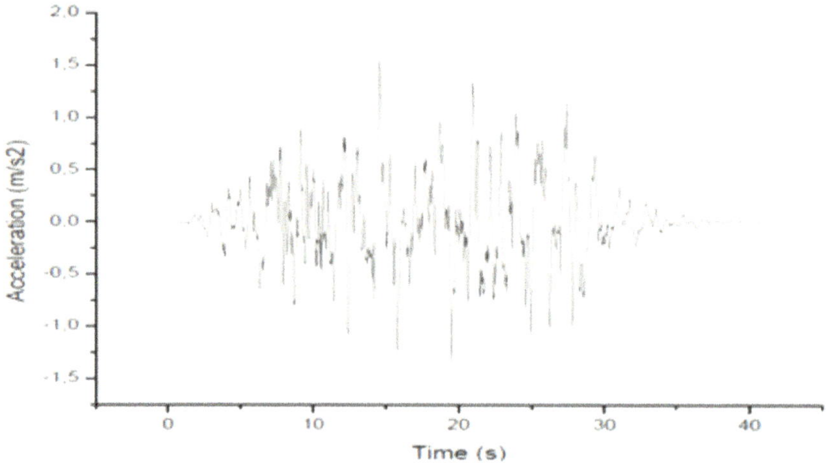

Figure 2: The accelerogram of El Centro California.

External stability analyses

To calculate the lateral earth pressure, Rankine's theory is used. For the dynamic case the calculation is based on the pseudo-static method.

Base Sliding-Static Case

The coefficient of Rankine active earth pressure is given by:

$$K_a = \tan^2(45 - \phi/2) \tag{4}$$

where K_a is the active earth pressure coefficient and f is the angle of soil internal friction. The static earth pressure is given by:

$$P_a = 0.5\, K_a\, \gamma_t\, H^2 \tag{5}$$

where P_a is the static earth pressure and γ_t and H are respectively the total unit weight of backfill soil and the height of the retaining wall.

The sliding resistance is provided by the friction between the concrete and the soil under the base of the wall. The frictional force between the concrete and the soil is computed using the following equation:

$$F_f = R_v \tan\delta \tag{6}$$

where F_f is the frictional force between the concrete and the soil, R_v is the vertical component of resultant force. F_r is the force opposing the sliding and:

$$F_r = F_f \tag{7}$$

$$R_v = W_{Total} \tag{8}$$

where W_{Total} is the resultant weight of the wall.

The safety factor against sliding is equal to:

$$F_s = F_r / P_a \tag{9}$$

Base Sliding- Dynamic Case

The pseudostatic earthquake force is given by [16]:

$$P_e = \frac{3\, a_{max}}{8g}\left(\gamma_t\, H^2\right) \tag{10}$$

where P_e is the pseudostatic earthquake force and a_{max} is the peak ground acceleration and:

$$P_{ae} = P_a + P_e \tag{11}$$

where P_{ae} is the sum of the static (P_a) and the pseudo-static earthquake force (P_e).

The safety factor for sliding using both pseudostatic method and Seed and Whitman (1970) analysis is given by:

$$F_s = \frac{N \tan \delta_1}{P_h + P_e} \tag{12}$$

where N is the sum of the weight of wall, the footing and the vertical component of the active earth pressure resultant force, δ_1 is the friction between the bottom of the foundation and the soil backfill, and P_h is the horizontal component of the active earth pressure resultant force with:

$$N = W + P_a \sin \delta \tag{13}$$

$$P_h = P_a \cos \delta \tag{14}$$

Overturning - Static Case

The overturning and stabilizing moments may be calculated by taking moments about point O. The safety factor against overturning is therefore:

$$F_{sr} = \frac{M_{st}}{M_r} \tag{15}$$

where M_{st} is the sum of moments that resist overturning and M_r is the sum of overturning moments.

Overturning - Dynamic case

The safety factor for overturning using Mononobe-Okabe method is given as:

$$F_s = \frac{W . a}{0.333 P_{ae} H \cos \delta - e P_{ae} \sin \delta} \tag{16}$$

where a the lateral distance from resultant weight W of wall and footing to toe of footing, H the height of the retaining wall and e the lateral distance from the vertical component of the active earth pressure resultant force to the toe of wall with:

$$P_v = P_a \sin \delta \tag{17}$$

where P_{ae} is the sum of the static (P_a) and the pseudostatic earthquake force (P_e).

If the conventional rule is applied, according to which the dynamic safety factor should not be less than 75%, the safety factor for sliding as well as for overturning should be in the range of 1.1 to 1.2.

NUMERICAL ANALYSES

The finite element method (FEM) has become the most popular method in both research and industrial numerical simulations. Several algorithms, with different computational costs can be implemented in ABAQUS [17], which is a commonly used software for finite element analysis. Understanding the nature, advantages and disadvantages of these algorithms is very helpful for choosing the right algorithm for a particular problem [18]. Comparison of implicit and explicit methods for ABAQUS in nonlinear problems has been reported by Rebelo in [19]. The unconditionally stable implicit method will encounter some difficulties when a complicated three dimensional model is considered. The reasons are as follows [18]: as the reduction of the time increment continues, the computational cost in the tangent stiffness matrix is dramatically increased and even causes divergence. Local instabilities cause force equilibrium to be difficult to achieve.

Explicit techniques are thus introduced to overcome the disadvantages of implicit methods. For the explicit method, the CPU cost is approximately proportional to the size of the finite element model and does not change as dramatically as the implicit method. The drawback of the explicit method is that it is conditionally stable. The stability limit for an explicit operator is that the maximum time increment must be less than a critical value of the smallest transition times for a dilatational wave to cross any element in the mesh. Secondly, the nature of the explicit method limits it to the analysis of short transient problems. If this method is used for quasi-static problems, the inertia effects must be small enough to be neglected. One way to assure this is to set the limit of the kinematic energy to be less than 5% of the strain energy. For dynamic problems, ABAQUS also offers some other methods such as a modal dynamic algorithm. However, only direct integration methods implicit dynamics and explicit methods are suitable for nonlinear problems. Most of the reported works on the comparison of implicit and explicit methods are on quasi-static nonlinear problems [18-20].

As it is well-known, the explicit integration operator is conditionally stable since the maximum stable time increment is limited by the minimum mesh size and the highest wave speed (Courant stability condition) given by [21-22]:

$$\Delta t \leq min\ \left(l_e / C_d\right)$$

where l_e is the characteristic element length (e.g. minimum element dimension) and C_d is the current effective, dilatational wave speed of the material. In other words, the time increment must be smaller than the minimum amount of time necessary for a wave to cross an element. The time step of $T = 2.2.10^{-4}$ s has been found to be sufficiently powerful in obtaining an effective approximate computations of the quasistatic response [23].

It should be mentioned that the values of damping for each material is regrouped in Table I. To get a clear comparison of the Von Mises stress distributions for the four studied cases, the Von Mises stress distributions are shown in Figures 3, 4, 5 and 6.

Retaining Wall Supporting Sand - Static Case:

According to the analytical computation, for this case, the safety factors for sliding as well as overturning are less than the default minimum values.]

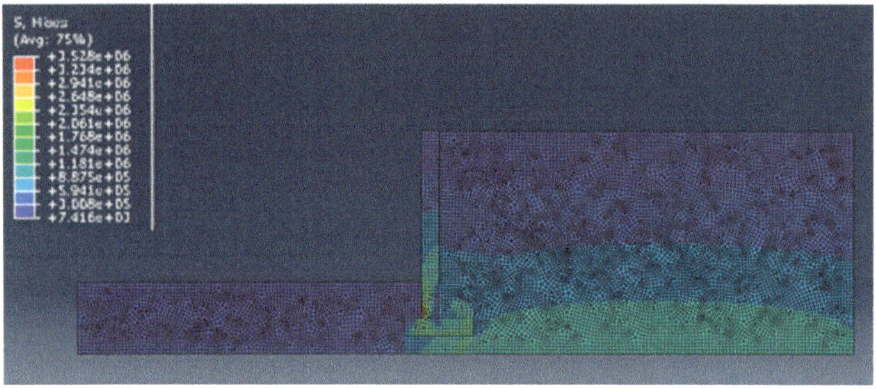

Figure 3: The Von Mises stress distribution for retaining wall suporting sand under static loading.

Von Mises stress distribution change in the range of 7.416– 3528 kPa (Figure 3). The lowest Von Mises stress values are observed near the ground level. Von Mises stress increases with depth. The higher Von Mises stress (3.5 MPa) is observed between the base slap and the stem of the retaining wall in the soil cover zone. Figure 7 presents U1 and U2 displacements under static loading for the two studied cases. Variations of U1 and U2 displacements of the wall with depth are presented in Figures 9 and 10. Case 1 and 2 make reference respectively to retaining wall supporting sand and retaining wall supporting reinforced sand. The displacement U1 is substantially constant with the depth (about 0.825 m). For U2 the maximum value is noted in the base slab and it's about 0.0065 m. The minimum value is noted in the top of the wall and it's about 0.0025 m.

Figure 5 presents the Von Mises stress distribution at 14 seconds (after 1023 time increments). For the second case, Von Mises stress distribution change in the range of 9.761 kPa to 13.95 MPa. The lowest Von Mises stress values (9.7 kPa) for this case are observed for backfill. The higher Von Mises stress (13.95 MPa) is observed in the wall between the base slap and the stem. The higher value of Von Mises stress for the dynamical case is 3.95 time higher than for the static case.

Figure 8 presents U1 and U2 displacements under static loading for the two studied cases. Variations of U1 and U2 displacements with depth are presented in Figure 11 and Figure 12. U1 displacement is about 0.70 m for all the depth. For U2 displacement, the maximum value is about 0.0053 m and the minimum value is about 0.002 m.

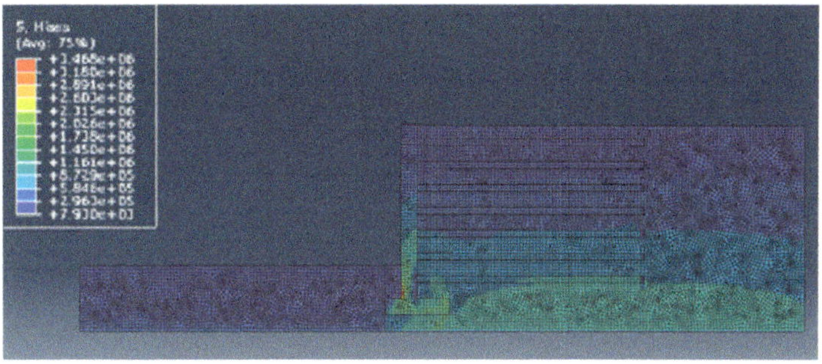

Figure 4: The Von Mises stress distribution for retaining wall suporting Tirereinforced sand under static loading.

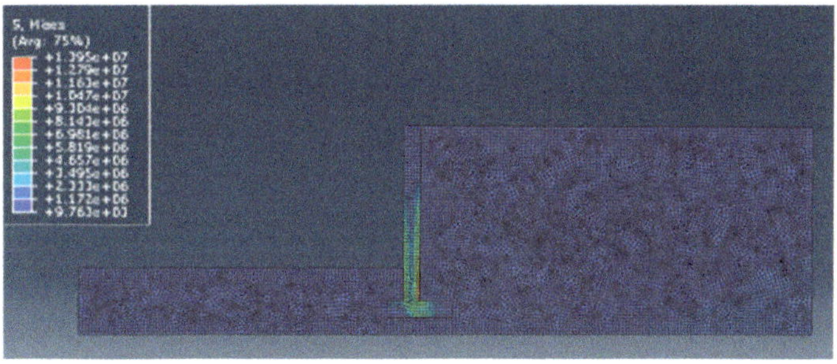

Figure 5: The Von Mises stress distribution for retaining wall suporting sand under seismic loading.

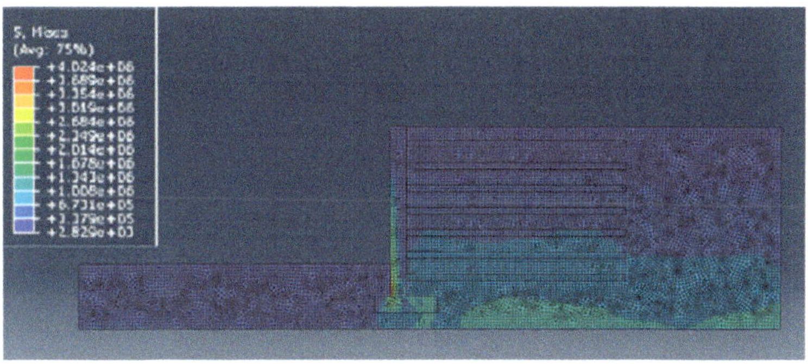

Figure 6: The Von Mises stress distribution for retaining wall suporting tirereinforced sand under seismic loading.

Retaining Wall Supporting Sand - Dynamic Case:

According to the analytical computation, for this case, the safety factors for sliding as well as overturning are greater than the default minimum values.

Retaining Wall Supporting Tire Reinforced Sand - Static Case:

According to the analytical computation, for this case, the safety factors for sliding as well as overturning are less than the default minimum values.

For the third case, the Von Mises stress distribution change in the range of 7.930 to 3465 kPa (Figure 4). The lowest Von Mises stress values are observed near the ground level. Von Mises stress increases with depth. The higher Von Mises stress is observed between the base slap and the stem of the retaining wall in the soil cover zone. There are not large differences between this case and the case of wall supporting sand for static loading.

Figure 7 presents U1 and U2 displacements under static loading for the two studied cases. Variations of U1 and U2 displacements with depth are presented in Figures 9 and 10. The value of U1 for this case is about 0.625 m, for U2 displacement the maximum value is about 0.005 m.

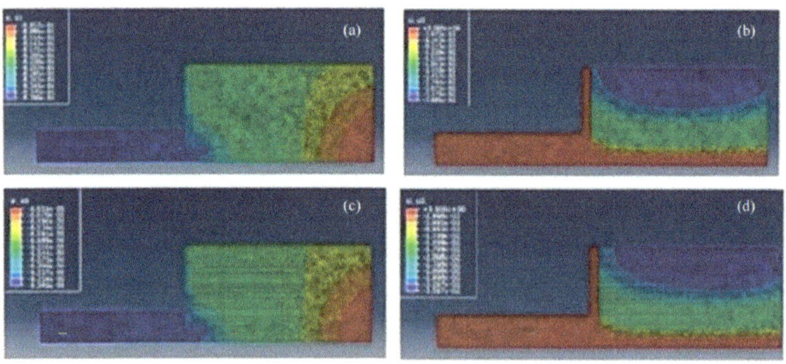

Figure 7: Displacements under static loading (a) U1 displacement for retaining wall suporting sand. (b) U2 displacement for retaining wall suporting sand. (c) U1 displacement for retaining wall suporting Tire-reinforced sand. (d) U2 displacement for retaining wall suporting Tirereinforced sand.

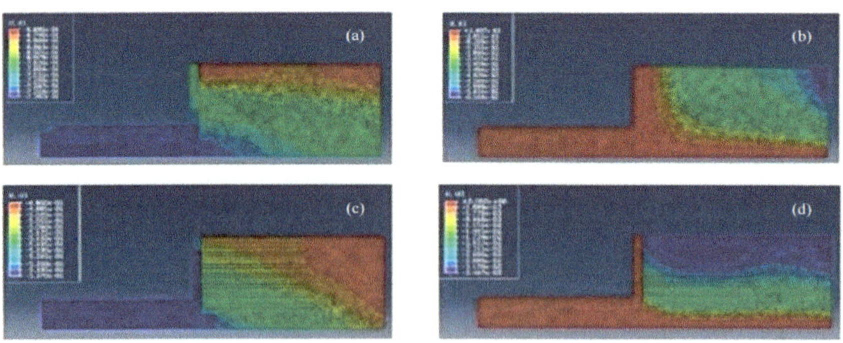

Figure 8: Displacements under seismic loading (a) U1 displacement for retaining wall suporting sand. (b) U2 displacement for retaining wall suporting sand. (c) U1 displacement for retaining wall suporting Tire-reinforced sand. (d) U2 displacement for retaining wall

Retaining Wall Supporting Tire Reinforced Sand - Dynamic Case:

Figure 6 presents the Von Mises stress distribution at 14 seconds (after 3869 time increments). For the fourth case, the Von Mises stress distribution change in the range of 2.829 kPa to 4.024 MPa. The lowest Von Mises stress values are observed near the ground level. Von Mises stress increases with depth. The higher Von Mises stress is observed between the base slap and the stem of the retaining wall in the soil cover zone. The higher value of Von Mises stress for this case is 3.46 times lower compared to the case of the retaining wall supporting sand with dynamic loading.

Figure 8 presents U1 and U2 displacements under static loading for the two studied cases. Variations of U1 and U2 displacements with depth are presented in Figures 11 and 12.

The U1 displacement is about 0.50 m and U2 displacement is about 0.0048 m.

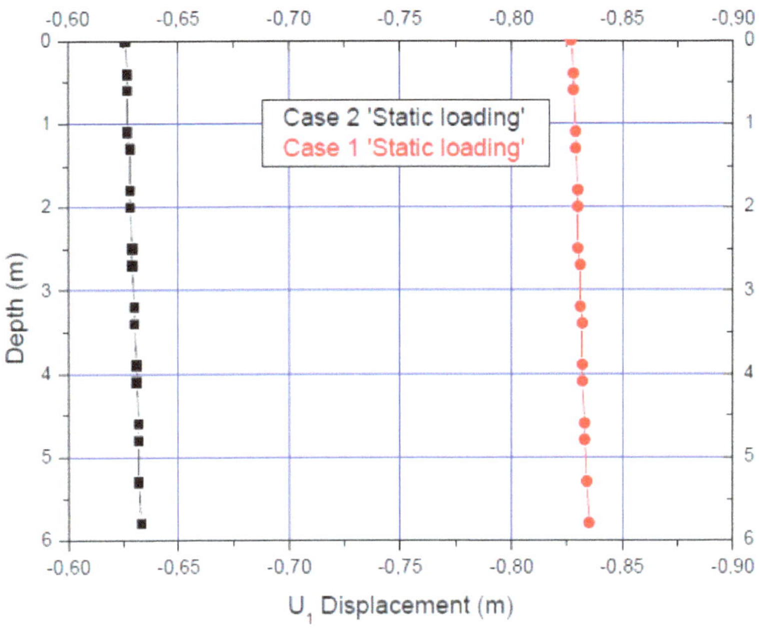

Figure 9: Variation of U1 wall displacement with depth for static loading.

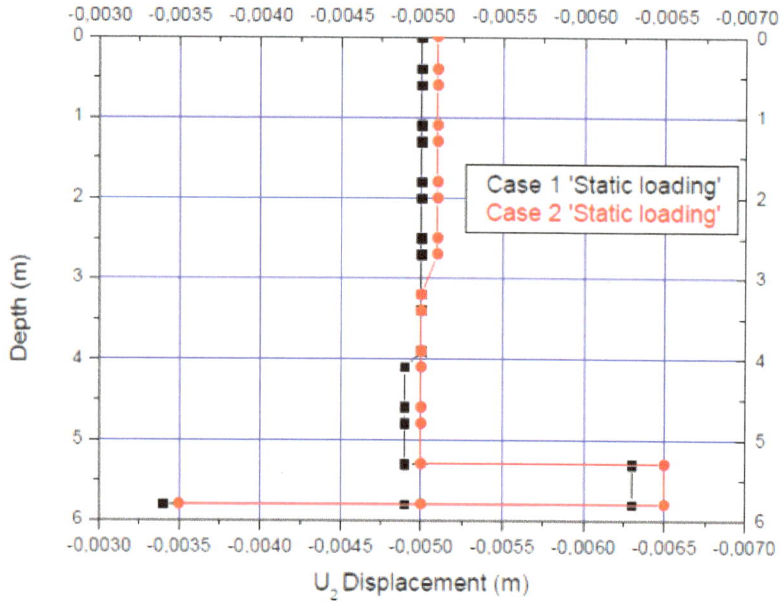

Figure 10: Variation of U2 wall displacement with depth for static loading.

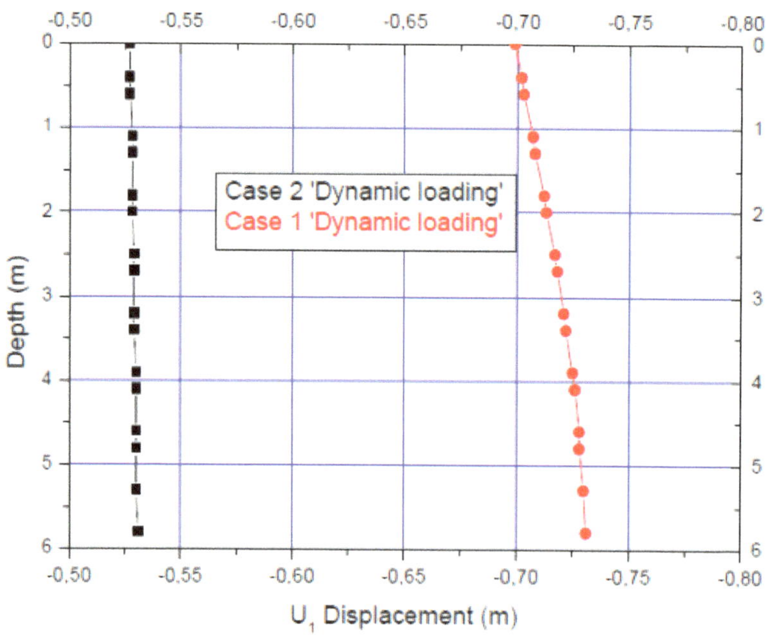

Figure 11: Variation of U1 wall displacement with depth for seismic loading.

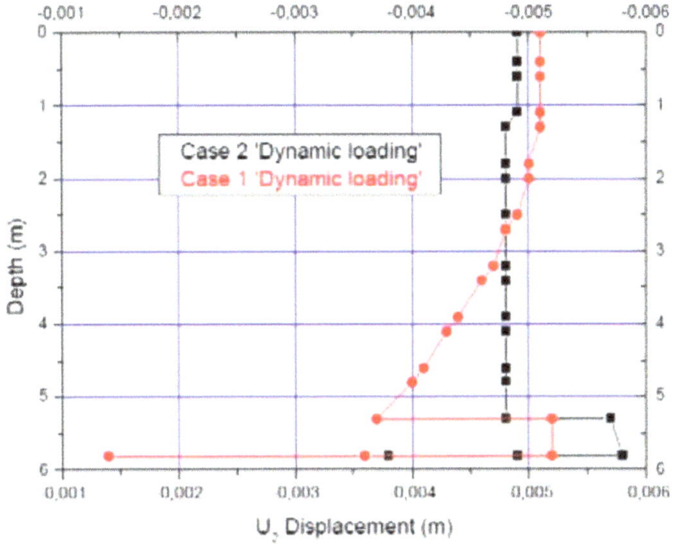

Figure 12: Variation of U2 wall displacement with depth for seismic loading.

CONCLUSION

Analytical analyses and implicit and explicit numerical methods are used to compare results of cantilever retaining wall supporting waste tire reinforced sand to the results of the same cantilever retaining wall supporting only sand, under seismic loading of El Centro type. The following conclusions are drawn:

- According to the analytical computation, the safety factors for sliding as well as for overturning are less than the default minimum values for all cases.
- For the retaining wall supporting sand case: the lowest Von Mises stress value is about 9.7 kPa whereas the highest value is about 13.95 MPa and is observed in the wall between the base slap and the stem. The higher value of Von Mises stress for the dynamical case is 3.95 times higher compared to the static case.
- For the retaining wall supporting waste tire reinforced sand under static loading case: the Von Mises stress distribution changes in the range of 7.930 to 3465 kPa and the lowest values are observed near the ground level. Under El Centro excitation, the higher Von Mises stress (4.024 MPa) is observed between the base slap and the stem of the retaining wall in the soil cover zone. This value is 3.46 times lower compared to the retaining wall supporting sand with dynamic loading.
- For U1 and U2 displacements, the variations with depth are similar for each type of loading, but the values for the case of retaining wall supporting waste tire reinforced sand are slightly lower compared to the ones of the case of retaining wall supporting only sand.

REFERENCES

1. C. Rothfuss, C. Y. Cha, "Utilization of a scrap tire-waste oil derived carbonous residue as an asphalt modifier", Polymer Recycling, Vol. 2, No. 3, pp. 201–212, 1996
2. C. Y. Wang, J. H. Cheng, H. P. Shih, J. W. Chang, "Ring columns as pier scour countermeasures", International Journal of Sediment Research, Vol. 26, No. 3, pp. 353-363, 2011
3. H. Trouzine, A. Asroun, N. Asroun, F. Belabdelouahab, N. T. Long, "Problématique des pneus usés en Algérie", Nature & Technologie, Vol. 5, pp. 28-35, 2011
4. in French.
5. N. T. Long, Le Pneusol: Recherches-Réalisations-Perspectives, Thèse de doctorat du LCPC, INSA Lyon, 1993
6. in French.
7. M. R. Hausmann, "Slope remediation", Proceedings of the Stability and Performance of Slopes and Embankments Conference, ASCE Geotechnical Special Publication, Vol. 2, pp. 1274–1317, 1990
8. V. K. Garga, V. O'Shaughnessy, "Tire-reinforced earthfill. Part 1: Construction of a test fill, performance, and retaining wall design", Canadian Geotechnical Journal, Vol. 37, No. 1, pp. 75–96, 2000
9. V. O'Shaughnessy, V. K. Garga. "Tire-reinforced earthfill. Part 2: Pullout behaviour and reinforced slope design", Canadian Geotechnical Journal, Vol. 37, No. 1, pp. 97–116, 2000
10. WYMCC, Reuse of worn tyres in civil engineering construction, Internal Report. West Yorkshire Metropolitan County Council, 1977
11. N. B. Romdhane, R. Nasri, M. Chaieb, N. T. Long, "Comportement dynamique du pneusol, Etude sur modèle reduit tridimensionnel, Essai sur table vibrante, Actes des 4eme Congres National AFPS, Saint-Rémy lés Chevreuse, France, 1996
12. In French.
13. Y. Guillard, N.T. Long, P. Ursat, J. C. Valleux, "Pneusol : Essais de vibrations, Actes des 2eme Congres National AFPS, France, 1989
14. In French.
15. S. L. Kramer, Geotechnical earthquake engineering, Englewood Cliffs, NJ: Prentice-Hall 1996
16. B. M. Basha, S. G. L. Babu, "Reliability assessment of internal stability of reinforced soil structures: A pseudo-dynamic approach", Soil Dynamics and Earthquake Engineering, Vol. 30, No. 5, pp. 336-353, 2010
17. D. Choudhury, S.M. Ahmad, "External stability of waterfront reinforced soil structures under seismic conditions using a pseudo-static approach", Geosynthetics International, Vol. 16, No.1, pp. 1–10, 2009

18. R. S. Steedman, X. Zeng, "The influence of phase on the calculation of pseudo-static earth pressure on a retaining wall", Geotechnique, Vol. 40, No. 1, pp. 103–112, 1990
19. A. K. Chopra, Dynamics of structures theory and applications to Earthquake Engineering, University of California at Berkeley, PrenticeHall, 1995
20. H. B. Seed, R. V. Whitman, "Design of earth retaining structures for dynamic loads", ASCE Specialty Conference on Lateral Stresses in the Ground and Design of Earth Retaining Structures, pp. 103-147, USA, 1970
21. ABAQUS User's Examples and Theory Manual, Version 5.7. Hibbitt, Karlsson & Sorensen Inc., 1998
22. J. S. Sun, K. H. Lee, H. P. Lee, "Comparison of implicit and explicit finite element methods for dynamic problems", Journal of Materials Processing Technology, Vol. 105, No. 1-2, pp. 110-118, 2000
23. N. Rebelo, J. C. Nagtegaal, L. M. Taylor, R. Passmann, "Comparison of implicit and explicit finite element methods in the simulation of metal forming processes", NUMINFORM 92, International Conference on Numerical Methods in Industrial Forming Processes, pp. 99-108, 1992
24. Hibbitt, Karlsson & Sorensen Inc., "Application of implicit and explicit finite element techniques to metal forming", Journal of Material Process Technology. Vol. 45, pp. 649-656, 1994
25. T. J. R. Hughes, The finite element method – linear static and dynamic element analysis, Englewood Cliffs, (NJ): Prentice-Hall, 1987
26. T. Belytschko, W. K. Liu, B. Moran, Nonlinear finite element for continua and structures, Chichester: John Wiley & Sons Ltd, 2000
27. L. Siad, M. Ould Ouali, A. Benabbes, "Comparison of explicit and implicit finite element simulations of void growth and coalescence in porous ductile materials", Materials and Design, Vol. 29, No. 2, pp. 319–329, 2008

13

CHAPTER

SUSTAINABILITY ASSESSMENT OF PRECAST ULTRA-HIGH PERFORMANCE FIBER REINFORCED CONCRETE (UHPFRC) CANTILEVER RETAINING WALLS

Behzad Nematollahi, M.R. Raizal Saifulnaz and Yen Lei Voo

Keywords: CO_2 emissions, Embodied Energy (EE), Global Warming Potential (GWP), precast cantilever retaining wall, sustainability, Ultra High Performance Fiber Reinforced Concrete (UHPFRC)

INTRODUCTION

Nowadays, there are many concerns worldwide over the environmental impacts such as CO_2 emission, Embodied Energy (EE) and Global Warming Potential (GWP) of our structural designs due to the increased public understanding of the devastating effects of global warming. It is believed that cement production contributes almost 5% of total CO_2 emissions in the world. As a rule of thumb, every 1 ton of cement produced would emit an averagely 0.81 ton of carbon dioxide (Huntzinger and Eatmon, 2009). Thus, it is required to minimize the environmental footprints of the structural designs. One of the solutions to achieve this goal is through Ultra-High Performance Fiber Reinforced Concrete (UHPFRC) technology that provides significant advantages in terms of aesthetic, workability, durability, ductility and sustainability. The UHPFRC technology supports the concept of sustainable construction through efficient use of material and optimization of the structural design. Further, the extremely durable nature of UHPFRC allows the UHPFRC structures to achieve much longer service life and almost negligible maintenance. This directly translates into the saving of cost and raw materials for re-construction (Nematollahi, 2012). UHPFRC also known as Steel Fiber Reinforced Reactive Powder Concrete (SFR-RPC) is one of the main breakthroughs in concrete technology in the mid 1990's. It was introduced by Richard and Cheyrezy

Citation: Behzad Nematollahi, M.R. Raizal Saifulnaz and Yen Lei Voo, Sustainability Assessment of Precast Ultra-High Performance Fiber Reinforced Concrete (UHPFRC) Cantilever Retaining Walls, ISSN: 2040-7459; e-ISSN: 2040-7467.

(1994, 1995) with compressive strength over 150 MPa and flexural strength over 30 MPa and remarkable improvement in durability compared to conventional concrete. To date, extensive research and development in the UHPFRC technology have been undertaken by numerous research groups and engineers worldwide. The material characteristics of UHPFRC have been studied in depth and its practical applications have been demonstrated in various countries throughout the world (Voo and Foster, 2009; Fehling et al., 2008; Graybeal, 2006; Schmidt et al., 2004). Most of the available literature of the UHPFRC structural members were focused on the experimental tests of the UHPFRC beams (especially prestressed beams) designed to fail in bending and/or shear (Voo et al., 2010; Voo and Foster, 2009; Voo et al., 2006). According to the most updated available literature on UHPFRC, there has been no study on the application of UHPFRC in conventional Reinforced Concrete (RC) cantilever retaining walls. Based on the enhanced mechanical properties of UHPFRC, it can offer numerous advantages in terms of aesthetic, workability, durability, ductility and sustainability when used in fabrication of precast concrete cantilever retaining walls.

This study presents the overview of a newly designed precast concrete cantilever retaining wall fabricated from UHPFRC as a sustainable alternative approach compared with the conventional precast RC cantilever retaining wall. Furthermore, the sustainability of the proposed UHPFRC wall is ascertained by comparing the Environmental Impact Calculations (EIC) of both walls with respects to material consumption, CO_2 emission, EE and 100-year Global Warming Potential (GWP).

MATERIALS AND METHODS

Analysis and Design of the Precast UHPFRC cantilever Retaining Wall

Figure 1 shows the proposed precast UHPFRC cantilever retaining wall with the dimensions of 2.5 m in height, 2 m in length and 2 m in width which supports a cohesionless horizontal backfill. It is assumed that the soil underneath the wall base and the soil in front of the wall have the same properties which are different from the properties of the backfill. Table 1 presents the soil properties used in this study according to soil characteristics given in Table 1 in BS 8002 (1994). In this study, the precast UHPFRC cantilever retaining wall was analyzed in accordance with BS EN 1997-1:2004 (2004) Eurocode 7 (EC7): Geotechnical design-Part 1: General rules requirements. At first, minimum dimensions of the wall were determined based on the stability and bearing pressure requirements given in EC7. The objective of the stability analysis is to ensure that the UHPFRC wall with the given dimensions is stable in terms of overturning and sliding under the action of the loads corresponding to the ULS (EQU) and the ULS (GEO), respectively. The bearing pressure analysis was also undertaken under the action of the loads corresponding to the ULS (GEO). Subsequently, the structural design of the stem, heel and toe of the wall were undertaken based on the first principles (equilibrium equations) in conjunction with the Japanese Society of Civil Engineers' Recommendations for Design and

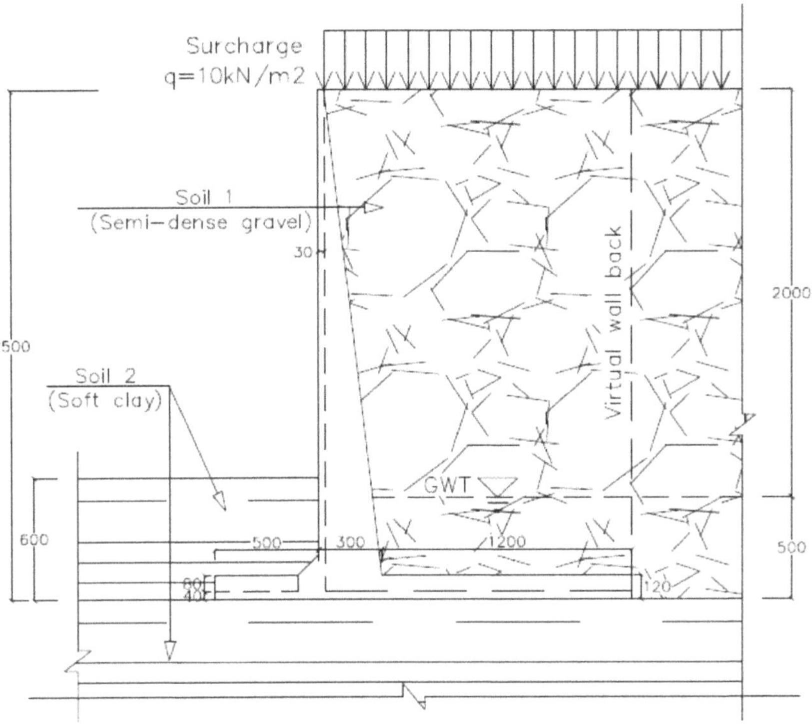

Figure 1: Soil and GWT conditions used in the analysis of the UHPFRC wall.

Construction of Ultra High Strength Fiber Reinforced Concrete Structures (Draft) (JSCE No. 9, 2006). Drawings of the precast UHPFRC cantilever retaining wall: The proposed precast UHPFRC cantilever retaining wall consists of two integrated thin panels which act as the base panel and the vertical wall panel. The 40 mm thick base panel was strengthened with two 80 mm thick by 100 mm wide steel reinforced stiffeners as shown in Fig. 2. The 30 mm thick vertical wall panel was also strengthened with two steel reinforced stiffeners as shown in Fig. 2 and 3. The action of possible hydrostatic pressure due to the percolating water during rain at the back face of the wall was reduced by arrangement of six weep holes with the diameter of 75 mm in the vertical wall panel as shown in Fig. 2. Besides, by provision of the weep holes, the possibility of seepage throughout the wall is decreased and water is less probable to reach and weaken the soil underneath the wall foundation (Mosley et al., 2007). Figure 2 and 3 show the detailed drawings of the precast UHPFRC cantilever retaining wall proposed in this study.

Analysis and Design of the Conventional Precast RC Cantilever Retaining Wall

As the benchmark of this study a corresponding 2.5 m high conventional RC wall as shown in Fig. 4 was also analyzed based on EC7 requirements with exactly the same soil,

Ground Water Table (GWT) and loading conditions as those of the precast UHPFRC wall; whereas the structural design of the conventional RC wall was undertaken based on the requirements given in BS EN 1992-1-1:2004 (2004), Eurocode 2 (EC2): Design of concrete structures- Part 1-1: General rules and rules for buildings. It should be noted that according to Murthy (2003); the minimum thickness of the wall stem and the minimum batter should be equal to 0.3 m and 1:48, respectively. In addition, similar to the UHPFRC wall, the minimum heel length required for development of the conjugate failure planes should be equal to 1.44 m. Thus, the conventional RC wall with the dimensions of 2.5 m in height, 1 m in length and 2.35 m in width was used in this study. Similar to the precast UHPFRC wall, the stability and bearing pressure analysis of the conventional RC wall with the above dimensions were undertaken according to EC7 requirements and all the requirements were met. Subsequently, the structural design of the wall was undertaken in accordance with EC2 requirements. Figure 5 shows the details of the bending reinforcement used in the corresponding conventional precast RC wall.

Table 1: Soil properties used in the analysis of the UHPFRC wall

	Soil type	
Soil properties	Semi dense gravel (soil 1)	Soft clay (soil 2)
Weight density (γ)	17 kN/m^3	17 kN/m^3
Saturated weight density (γ_{sat})	20.5 kN/m^3	17 kN/m^3
Effective cohesion (c')	0 (cohesionless)	22 (cohesive)
Characteristic angle of shearing resistance ($\varnothing'_k$)	30°	35°
Critical state angle of shearing resistance ($\varnothing'_{cv}$)	30°	35°

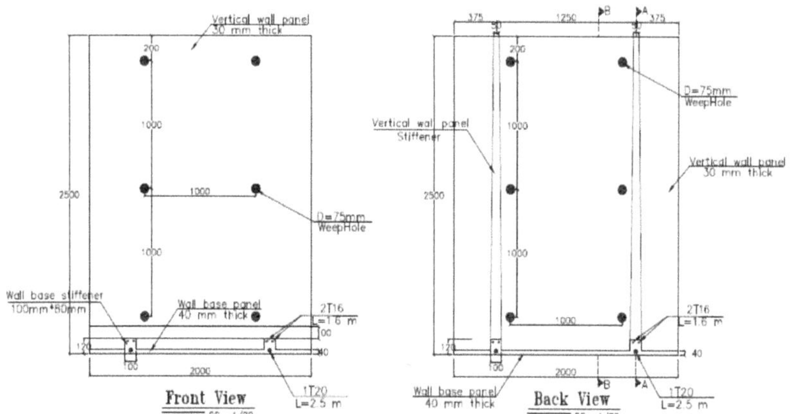

Figure 2: Front view and back view of the precast UHPFRC cantilever retaining wall.

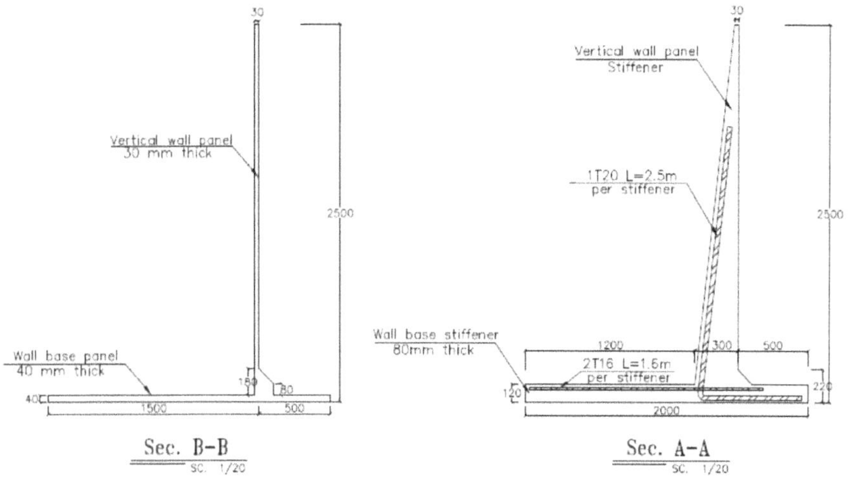

Figure 3: Sections A-A and B-B of the precast UHPFRC cantilever retaining wall.

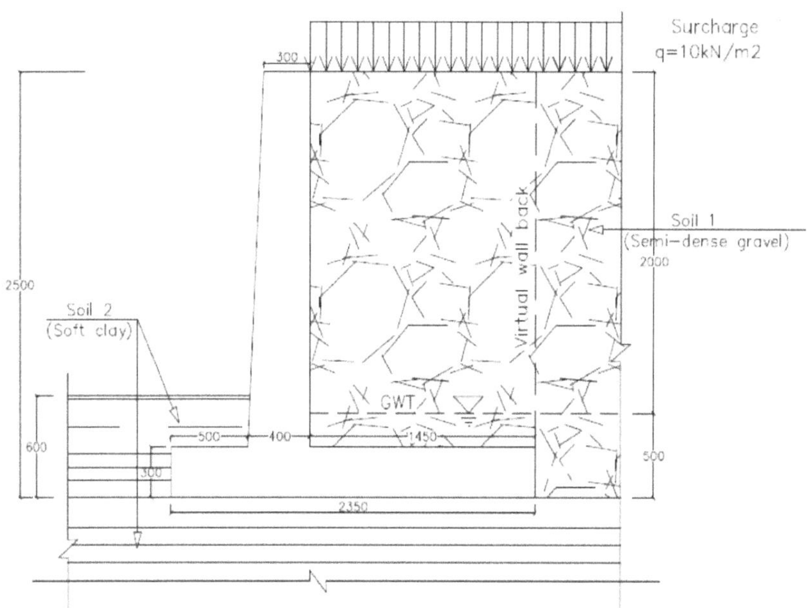

Figure 4: Dimensions, soil and GWT conditions used in the analysis of the RC wall as the benchmark of this study.

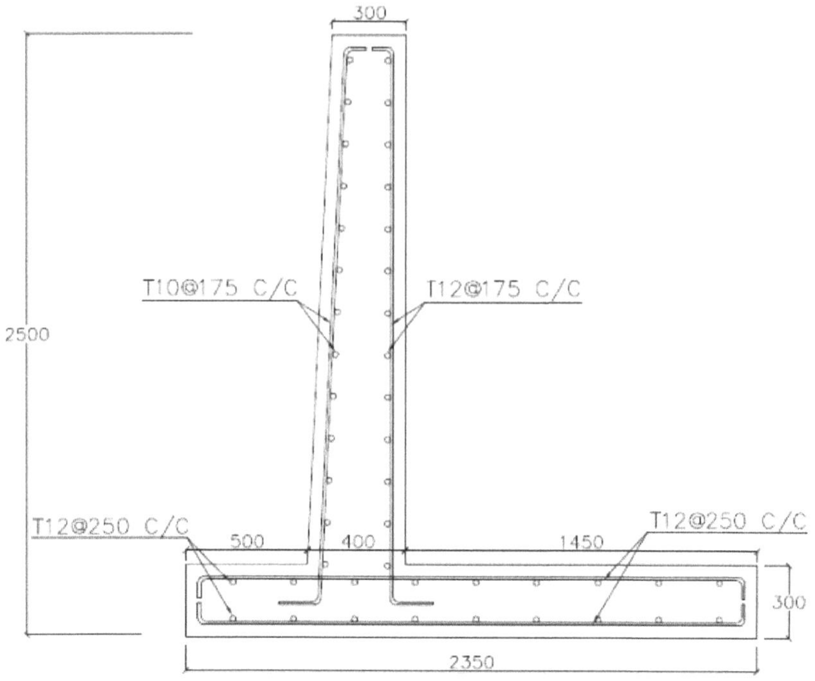

Figure 5: Details of the bending reinforcement used in the RC wall in STR (ULS).

Table 2: Environmental data used for the EIC

	Units	DURA-UHPFRC (1.5% steel fiber)*	Grade-30 (15% PFA)	Reinforcement
Density	kg/m^3	2350	2350	7840
EE	GJ/m^3	6.814	1.73	185.8
CO_2	kg/m^3	982	297.50	17123
NO_x	kg/m^3	4.860	1.66	55.4
CH_4	kg/m^3	0.760	0.12	30.7
100-year GWP	kg CO_2 eq./m^3	2449	795	34392

*: Environmental values include steel fiber contribution; (Voo and Foster, 2010)

Environmental Impact Calculations (EIC)

Environmental Impact Calculations (EIC) of the precast UHPFRC cantilever retaining wall was undertaken to ascertain that the proposed precast UHPFRC cantilever retaining wall as a sustainable alternative solution to the conventional precast RC cantilever retaining walls supports the concept of sustainable construction. The environmental data needed for EIC is presented in Table 2. This table summarizes the values of equivalent CO_2 content, the EE and 100-year GWP of Grade-30 concrete, of DURA®-UHPFRC with 1.5% of steel fibers and of steel reinforcement. The CO_2 emission, EE and GWP values for the production of concrete and steel are obtained from the study of Voo and Foster (2010) and Struble and Godfrey (2004) and are modified as required. GWP is a measure to estimate the contribution of a particular mass of greenhouse gas to global warming over an agreed time intervals (Elrod, 1999). According to Voo and Foster (2010), for simplicity the following formula can be used to measure the 100-year GWP with a unit of kilograms of CO_2 equivalent per kilogram of material (kg CO_2 eq./kg):

Table 3: Material quantities and EIC of the two cantilever retaining wall designs

Design method		UHPFRC (1.5% steel fiber) (m^3)	Grade 30 (15% PFA) (m^3)	Reinforcement (kg)	Total
Conventional RC wall	Grade 30 concrete	0	1.47	0	-
	Steel bars	0	0	70.80	-
	Mass of materials used; (kg)	0	3454.50	70.80	3525.3
	EE; (GJ)	0	2.54	1.68	4.2
	CO_2; (kg)	0	437.30	154.30	591.6
	100-yr GWP; (kg CO_2 eq.)	0	1168.70	310.80	1479.5
UHPFRC wall	UHPFRC concrete	0.208	0	0	-
	Steel bars	0	0	14.65	-
	Mass of materials used; (kg)	488.800	0	14.65	503.5
	EE; (GJ)	1.420	0	0.35	1.8
	CO_2; (kg)	204.300	0	31.90	236.2
	100-year GWP; (kg CO_2 eq.)	509.400	0	64.30	573.7

$$100\text{-year GWP} = CO_2 + 298\ NO_x + 25\ CH_4 \tag{1}$$

RESULTS AND DISCUSSION

Table 3 summarizes the material quantities and EIC of the two concrete cantilever retaining wall designs (i.e., the conventional precast RC wall and the newly designed precast UHPFRC wall). The amount of EE, CO_2 emissions and 100-year GWP are derived from multiplying the total amount of each material used in each of the precast wall specimens by their corresponding unit value of the environmental data given in Table 2. It should be pointed out that foundation of the wall and transportation costs are excluded in the calculation of the material quantities of the walls i.e., the material quantities shown in Table 3 are just based on the precast wall specimens. According to this table, the precast UHPFRC wall weighs 503.5 kg/m length of the wall; while, the conventional precast RC wall weighs 3525.3 kg/m length of the wall. In other words, the precast UHPFRC wall is seven times lighter than the conventional precast RC wall! In addition, lighter weight of the precast UHPFRC wall will result in a smaller foundation which provides additional savings. Furthermore, due to the lighter structure of the precast UHPFRC wall its standard length is 2 m; whereas the standard length of the conventional precast RC wall is usually 1 m. Hence, the precast UHPFRC wall does not need heavy lifting and installation machineries compared to the conventional precast RC wall which leads to a reduction in transportation costs and installation time. With respect to reinforcement, the precast UHPFRC wall eliminates the use of secondary reinforcements and crack control bars needed in the stem and the base of the conventional precast RC wall. In addition, it also removes the need of reinforcement required in the compression face of the wall stem. Thus, the total weight of the reinforcement needed is many times less than the conventional precast RC wall.

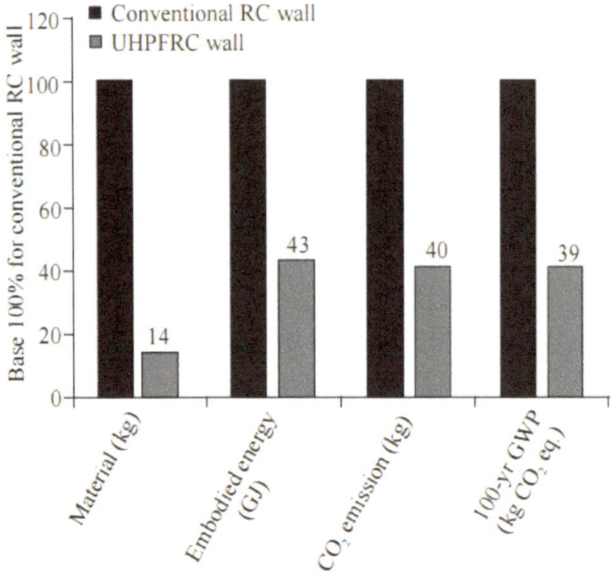

Figure 6: EIC results of the two cantilever retaining wall designs.

Figure 6 shows a comparison of the EIC results of the two cantilever retaining wall designs based on 100% for the conventional RC wall. As can be seen in this figure, regarding material consumption, the proposed precast UHPFRC wall consumed 86% less material than the conventional precast RC wall. Further, with regards to environmental impact, the precast UHPFRC wall has 57% less Embodied Energy (EE) and 60% less CO2 emissions compared with the precast RC wall. In addition, with respect to the 100-year GWP, the precast UHPFRC wall offers a decrease of 61% over that of the conventional precast RC wall. It is necessary to point out that only the savings at the stage of the wall design have been considered in this study. Considering the lighter weight of the precast UHPFRC wall will result in a smaller foundation and lower transportation costs, thereby additional savings will be obtained.

CONCLUSION

This study evaluates the environmental impacts of a newly designed precast UHPFRC cantilever retaining wall. According to the results of EIC, in terms of raw material consumption the precast UHPFRC wall gives immediate saving of 86% compared against the conventional precast RC wall. In other words, the precast UHPFRC wall is seven times lighter than the conventional precast RC wall. In addition, in term of environmental impact indexes the precast UHPFRC wall is confirmed to be much more sustainable as its EE and CO2 emission values are approximately 57 and 60% less than those of the conventional precast RC wall, respectively. Moreover, in terms of the 100-year GWP, a decrease of 61% is achieved by using the precast UHPFRC wall over that of the conventional precast RC wall. It is necessary to indicate that these savings are only obtained in the stage of the wall design and further savings will be achieved if the smaller foundation and reduced transportation costs are considered resulted by the lighter weight of the precast UHPFRC wall. In summary, the precast UHPFRC cantilever retaining wall proposed in this study is a green structural member that supports the concepts of sustainable construction which can be used as an alternative sustainable approach compared to the conventional precast RC cantilever retaining walls.

REFERENCES

1. BS 8002, 1994. Code of Practice for Earth Retaining Structures. British Standards, UK.BS EN 1992-1-1:2004, 2004. Eurocode 2:
2. Design of Concrete Structures- Part 1-1: General Rules andRules for Buildings. British Standards, UK. BS EN 1997-1:2004, 2004. Eurocode 7: Geotechnical
3. Design-Part 1: General Rules. British Standards, UK.
4. Elrod, M., 1999. Greenhouse warming potential model. J. Chem. Educ., 76: 1702-1705.
5. Fehling, E., M. Schmidt and S. Sturwald, 2008. Ultra High Performance Concrete (UHPC). Proceedingsof the 2nd International Symposium on Ultra High Performance Concrete. Kassel, Germany.
6. Graybeal, B., 2006. Material Property Characterization of Ultra High-performance Concrete. US Department of Transportation, Federal Highway Administration, Publication No: FHWA-HRT- 6-103.
7. Huntzinger, D.N. and T.D. Eatmon, 2009. A life-cycle assessment of Portland cement manufacturing: Comparing the traditional process with alternative technologies. J. Clean. Prod., 17(7): 668-675.
8. JSCE No. 9, 2006. Recommendations for design and construction of ultra high strength fiber reinforced concrete structures (Draft). JSCE Guideline for Concrete, No. 9: Concrete Committee of JapanSociety of Civil Engineers (JSCE).
9. Mosley, W.H., J.H. Bungey and R. Hulse, 2007. Reinforced Concrete Design to Eurocode. 6 Edn., Palgrave MacMillan, pp: 434.
10. Murthy, V., 2003. Geotechnical Engineering: Principles and Practices of Soil Mechanics and Foundation
11. Engineering. Dekker, New York.
12. Nematollahi, B., 2012. Structural behavior and environmental impacts of precast ultra-high performance ductile concrete cantilever retaining walls. M.Sc. Thesis, Universiti Putra Malaysia (UPM), Serdang, Malaysia.
13. Richard, P. and M. Cheyrezy, 1994. Reactive powder concretes with high ductility and 200-800 MPa compressive strength. Int. Concrete Res. Inform. Portal, 144: 507-507.
14. Richard, P. and M. Cheyrezy, 1995. Composition of reactive powder concretes. Cement Concrete Res.,25(7): 1501-1511.

15. Schmidt, M., E. Fehling and C. Geisenhansluke, 2004.Ultra High Performance Concrete (UHPC).Proceedings of the International Symposium onUltra High Performance Concrete. KasselUniversity Press, Kassel, Germany.
16. Struble, J. and J. Godfrey, 2004. How sustainable isconcrete? Proceedings of the InternationalWorkshop on Sustainable Development andConcrete Technology. Beijing.
17. Voo, Y.L. and S.J. Foster, 2009. Reactive Powder Concrete: Analysis and Design of RPC Girders. Lambert Academic Publishing, Saarbrücken, Germany.
18. Voo, Y.L. and S.J. Foster, 2010. Characteristics of ultra-high performance 'ductile'concrete and its impact on sustainable construction. IES J. A: Civ. Struct. Eng., 3(3): 168-187.
19. Voo, Y.L., S.J. Foster and R.I. Gilbert, 2006. Shear strength of fiber reinforced reactive powder concrete prestressed girders without stirrups. J. Adv. Concrete Technol., 4(1): 123-132.
20. Voo, Y.L., W.K. Poon and S.J. Foster, 2010. Shear strength of steel fiber-reinforced ultra-high performance concrete beams without stirrups. J. Struct. Eng., 136(11): 1393-1400.

14

CHAPTER

A NOTE ON THE ANALYSIS AND DESIGN OF CANTILEVER RETAINING WALLS USING THE CECP2 AND BS8002 METHODS

Shodolapo Oluyemi Franklin and Abiodun Luqman Olopade

Keywords: Cantilever, wall, stability, sliding, overturning, bearing, pressure, factor, safety

INTRODUCTION

Earth retaining structures are ubiquitous in the man-made environment. However, notwithstanding, their use has increased enormously in recent decades due to the great advance in technological, social and economic developments that have occurred. Reinforced concrete cantilever retaining walls are considered to be more economical in comparison to their traditional mass concrete or masonry gravity wall counterparts. This is because unlike the latter whose material is used solely for its dead weight, in cantilever walls the backfill itself contributes significantly towards the required dead weight. As a consequence the concrete stem can be made relatively more slender thus, in addition, enhancing the visual appearance and aesthetics of the structure. Corresponding to the greatly increased usage noted above has been an increased development in analysis and design methods with respect to earth retaining structures of various types. An excellent summary of some of the significant developments is provided by Craig [1]. The CECP2 code [2] represents one of the very popular conventional methods which has been applied reasonably successfully to retaining structures. The BS8110 design code [3] for reinforced concrete structures has also been utilized in conjunction with CECP2in the quest for satisfactory analysis and design solutions. Despite the previous comments however it has been observed that the CECP2 approach possesses a number of shortcomings and might not yield reliable nor accurate results in several cases [4].

Citation: Shodolapo Oluyemi Franklin and Abiodun Luqman Olopade, A Note On The Analysis And Design Of Cantilever Retaining Walls Using The CECP2 and BS8002 Methods, ISSN: 2076-734X (Print); 2076-7366 (Online).

An alternative and more recent method represented by the BS8002 code [5] embodies limit state principles to an extent and marks a sharp deviation from the conventional approach of CECP2. However the existence of two widely used and apparently conflicting methods has generated further observations as well as controversy in respect of appropriate procedures for analysis and design of earth retaining structures. As a result of the foregoing developments the present investigation was embarked upon. The primary objective was to compare and contrast the results of analysis and design of a typical reinforced concrete cantilever retaining wall using both the CECP2 and BS8002 codes. The approach utilized has been mainly theoretical and case studies and investigations of existing walls have been avoided. However the soil design parameters that have been adopted for this purpose are typical of those employed locally [6].

METHODOLOGY

The design of a reinforced concrete cantilever retaining wall according to CECP2 in conjunction with BS8110 involves several considerations. Firstly stability analysis is based on the ultimate limit state, the critical condition being when a maximum horizontal force acts with a minimum vertical load. In order to guard against a stability failure, a safety factor γ_f 1.6 = will be applied. Secondly the bearing pressure analysis is based on the serviceability limit state. In the BS8002 methodology, a fraction of the peak strength of the soil is used. Although limit state philosophy is adopted, partial safety factors are not employed. The design earth pressures are based on those which occur at serviceability. The design including stability against sliding and overturning is carried out using a mobilization factor M. A minimum value of M = 1.2 is specified for both angle of shearing resistance ϕ and cohesion c, for drained conditions in the soil. In addition the design value of the angle of wall friction δ will be taken as $\tan^{-1}$ ($0.75\tan\phi$). Furthermore a compulsory 0.5 m over-dig level is imposed in addition to a minimum 10 KN/m^2 surcharge. No further factors of safety are necessary because according to BS8002 the design is based on the worst case scenario. The comparative design is done using the cantilever retaining wall shown in Figure 1. The wall supports a granular material of unit weight 19.6 KN/m^3 and the allowable bearing pressure is 110 KN/m^2. The height of the soil to be retained is 4.5 m and the thickness of the base slab is 0.4 m. The width of the base slab is 3.4 m and the toe projection is set at 0.8 m. These dimensions represent that of a typical cantilever retaining wall. For the present study an angle of shearing resistance $\phi = 30^\circ$ has been adopted. With BS8002 it is necessary to consider soil properties as stipulated in the code in order to determine the design ϕ. The latter depends on soil properties such as angularity of the soil particles, the grading of the particles and the results of standard penetration tests, all expressed in degrees. Analysis was carried out by varying the soil properties and choosing the appropriate ϕ value, and subsequently the design ϕ and other design parameters. The maximum or peak ϕ is given by

$$\phi_{max} = \phi + A^o + B^o + C^o \tag{1}$$

where A, B and C depend on the angularity of soil particles, the grading and the standard penetration test results in that order. The design ϕ is given by

$$\text{design } \phi = \tan^{-1}[(\tan\phi_{max} / M)] \quad (2)$$

where M is the mobilization factor. In all cases the value of C is taken as 2° for sand/gravel soil, implying less than 20 blows in the standard penetration test. In line with BS8002 methodology, the shear strength is taken as the lesser of the peak strength divided by the mobilization factor or the critical state strength.

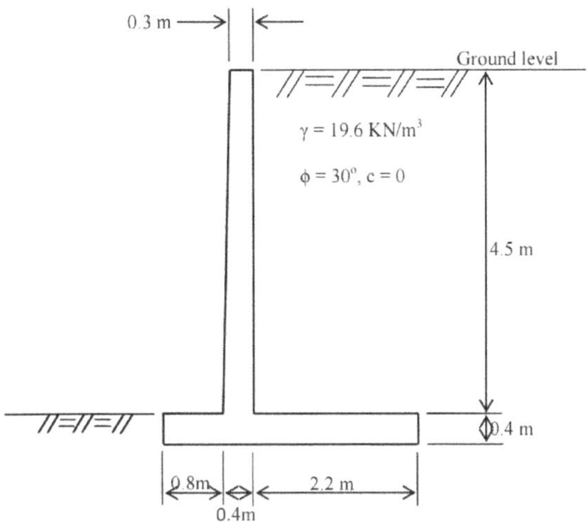

Figure 1: Schematic of reinforced concrete cantilever retaining wall

RESULTS AND DISCUSSION

The results of the analysis and design of the cantilever retaining wall in granular soil for both the conventional CECP2 code used in conjunction with BS8110, and the BS8002 approach, are shown in Table 1. With reference to the design parameters in respect of both codes, every attempt has been made to ensure uniformity in order to assist the comparative study. It is obvious from Table 1 that both design methods give varying results, a situation attributed to the differences in design procedures.

Examination of Table 1 reveals that for both codes in all cases, stability against sliding and overturning is achieved. More specifically however, using the conventional CECP2 approach coupled with the safety factors stipulated by BS8110, stability against sliding is achieved only with the provision of a heel beam. In comparison if the BS8002 design method is used, no allowance for a heel beam need be made despite the code's insistence on the most onerous loading condition. Interestingly for the bearing pressure analysis using the conventional CECP2 method, maximum bearing pressure for this specific case is less than the allowable bearing pressure. In contrast for the BS8002 approach, maximum bearing pressure is less than the allowable value provided the design φ is very high, a situation that corresponds to the use of angular soil particles. For values of design φ less than 33° approximately, maximum bearing pressures always exceed the allowable values. Since the value of the maximum bearing pressure is primarily dependent only on the backfill and load above the base, and not on the soil below the base slab, it can be inferred that the use of angular soil particles for the present case should result in a more economical design for the cantilever wall. The relationships between the design parameters A, B and ϕ and the factors of safety against sliding and overturning are shown in Figures 2 and 3 respectively. The ratio (A+B)/ϕ is a dimensionless parameter, and the graphs demonstrate that stability of the wall in respect of sliding and overturning increases with increase in this dimensionless parameter.

Table 1: Results of cantilever retaining wall design in respect of CECP2/BS8110 and BS8002 codes

Soil properties	Design parameters	Stability against overturning (*)	Stability against sliding (**)	Bearing pressure analysis (***)
Conventional method using CECP2 in conjunction with BS8110				
No provision for angularity or the grading of particles	Design $\phi = 30°$ Design $\delta = 20°$	540/172.2 = 3.14 Satisfied	107.1/105.4 = 1.02 Satisfied with a heel beam provided	110/86.9 = 1.27 satisfied
BS8002 method				
Rounded particles A = 0° Uniform grading B = 0°, C = 2°.	$\phi_{max} = 32°$ Design $\phi = 27.5°$ Design $\delta = 21.3°$ $(A + B)/\phi = 0$	643.9/204.7 = 3.15 Satisfied	121.8/105.3 = 1.16 Satisfied	110.0/132.1 = 0.83 Not satisfied
Rounded particles A = 0° Moderate grading B = 2°, C = 2°	$\phi_{max} = 34°$ Design $\phi = 29.3°$ Design $\delta = 22.8°$ $(A + B)/\phi = 0.07$	643.9/184.9 = 3.48 Satisfied	131.6/95.10 = 1.38 Satisfied	110.0/122.5 = 0.90 Not satisfied
Sub-angular particles A = 2° Moderate grading B = 2°, C = 2°	$\phi_{max} = 36°$ Design $\phi = 31.2°$ Design $\delta = 24.4°$ $(A + B)/\phi = 0.13$	643.9/176.3 = 3.65 Satisfied	141.6/88.3 = 1.60 Satisfied	110.0/116.1 = 0.95 Not satisfied
Angular particles A = 4° Moderate grading B = 2°, C = 2°	$\phi_{max} = 38°$ Design $\phi = 33.1°$ Design $\delta = 26.1°$ $(A + B)/\phi = 0.18$	643.9/158.4 = 4.07 Satisfied	152.9/81.5 = 1.88 Satisfied	110.0/109.6 = 1.00 Satisfied

(*) Stability = Restoring moment/Overturning moment
(**) Stability = Frictional resisting force/Sliding force
(***) Stability = Allowable bearing pressure/Maximum bearing pressure

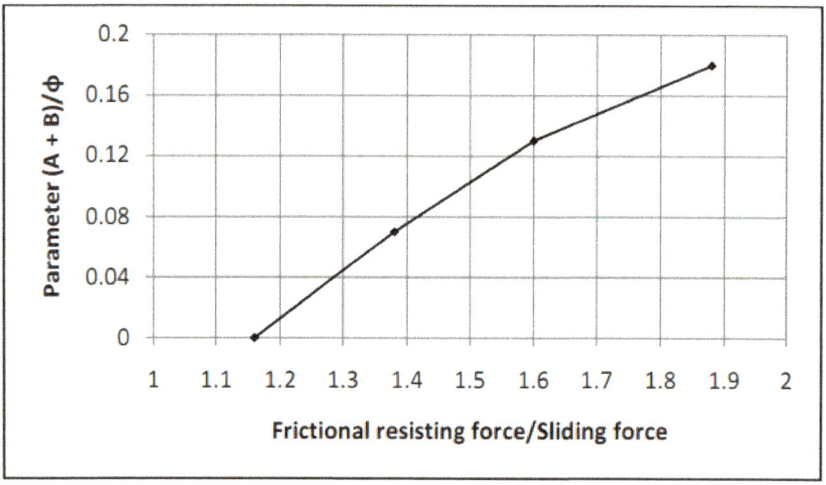

Figure 2: Relationship between design parameters (A, B, and ϕ) and factor of safety against sliding

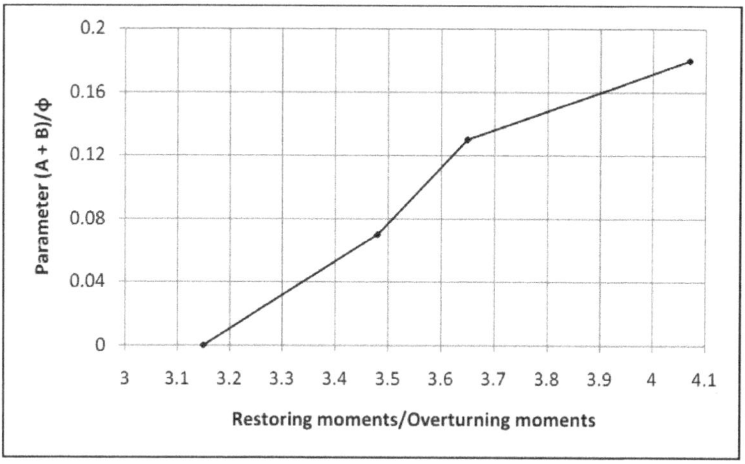

Figure 3: Relationship between design parameters (A, B, and ϕ) and factor of safety against overturning

CONCLUSIONS

In the present study, the CECP2 and BS8002 methods have been applied to the analysis and design of a typical cantilever retaining wall supporting a granular backfill. Representative values of soil design parameters have been employed for this purpose. From the results presented, the following conclusions have been reached.

a) With the CECP2 method used in tandem with BS8110, stability against sliding is achieved only with the provision of a heel beam. In contrast with the BS8002 approach, despite its insistence on the most onerous loading conditions, the need for a heel beam is found to be unnecessary.

b) For the specific design case studied, maximum bearing pressures are below the allowable value when the CECP2 method is used. On the other hand with the BS8002 approach, the bearing pressure criterion is satisfied only when the design angle of shearing resistance ϕ is high, a condition which corresponds to the use of angular soil particles.

c) For the BS8002 method, the stability of the wall in relation to sliding and overturning increases with increase in the dimensionless parameter $(A + B)/\phi$.

REFERENCES

1. Craig, R.F., 2004, Craig's Soil Mechanics, 7th Edition, Taylor and Francis, London, pp. 161–226.

2. Civil Engineering Codes of Practice Joint Committee, 1951, Earth Retaining Structures: CECP2, Instn. Struct. Engs., London.

3. British Standards Institution, 1985, Structural Use of Concrete - Part 1: Code of Practice for Design and Construction (BS8110), British Standards Institution, London.
4. Burland, J.B., Potts, D.M. and Walsh, N.M., 1981, The Overall Stability of Free and Propped Cantilever Retaining Walls, Ground Engineering, Vol. 14, No. 5, pp. 28–38.
5. British Standards Institution, 1994, Code of Practice for Earth Retaining Structures (BS8002), British Standards Institution, London.
6. Franklin, S.O. and Olopade, A.L., 2012, A Commentary on the Use of the CECP2 and BS8002 Methods for the Analysis and Design of Cantilever Sheet Piles, Journal of Basic and Applied Scientific Research, TextRoad Publication, Vol. 2, No. 2, pp. 1222–1229.

15
CHAPTER

SOME REMARKS ON THE APPLICATION OF THE CECP2 AND BS8002 METHODS FOR THE ANALYSIS AND DESIGN OF ANCHORED SHEET PILE WALLS

Shodolapo Oluyemi Franklin, and Abiodun Luqman Olopade

Keywords: Anchored, pile, cohesive, cohesionless, embedment, depth, factor, safety

INTRODUCTION

Retaining walls have found increasing use in construction in recent decades due to the greatly increased need for urban, economic and social development all over the world. Anchored sheet pile walls or, more succinctly, anchored bulkheads serve the same function as retaining walls but, unlike the latter whose weight is a considerable fraction of the weight of the sliding wedge, bulkheads generally comprise a single row of comparatively light sheet piles driven to earth at the lower ends but with the upper ends anchored by tie rods. The tie rods are held securely by anchors buried in the backfill some distance from the sheet pile. The use of anchor rods will generally reduce the depth of penetration, lateral deflection and the bending moments in the pile [1]. On account of the ever increasing demand for these structures several notable developments in the analysis and design of anchored and cantilever sheet piles have taken place in recent decades [2–12]. An overview of the more significant developments is provided by Craig [13] and no further detailed comments are deemed necessary here. One of the more popular conventional methods which have been applied in a number of variant forms is that represented by the CECP2 code [3]. Although the method has often provided satisfactory solutions in several cases it has been found not to be totally accurate and to possess some shortcomings [6]. An alternative and relatively more recent approach contained in the BS8002 code [14] is based on limit state design philosophy and represents a radical shift from the conventional method of CECP2. However this

Citation: Shodolapo Oluyemi Franklin & Abiodun Luqman Olopade, Some Remarks On The Application Of The Cecp2 And Bs8002 Methods For The Analysis And Design Of Anchored Sheet Pile Walls, ISSN: 2076-734X (Print); 2076-7366 (Online).

dichotomy has generated a lot of controversy in the geotechnical community in respect of the analysis and design of earth retaining structures. It is as a consequence of the above that the present study was embarked upon. The main objective was to compare and contrast the results of analysis and design of anchored sheet pile walls using the CECP2 and BS8002 methods. An earlier study [15] was devoted entirely to cantilever sheet piles. Case studies and investigations of existing walls have not been attempted here and for the present study the approach employed has been mainly theoretical. The values of soil design parameters utilized were generally assumed although such values could be described as good representations of those obtained in local practice.

METHODOLOGY

There are two classical methods currently in use for the design of anchored sheet pile walls, the free- earth and the fixed-earth approach. However the latter is adopted in the present study on account of its greater simplicity. It is assumed that the sheet pile is rigid and will rotate about the anchor rod level. Passive pressure develops in front of the pile while active pressure develops behind the embedded wall. Failure occurs by rotation about the fixed anchor rod. In relation to the conventional CECP2 method and its variants, once the embedment depth is calculated its value is increased by approximately 20% to 40%, or alternatively, a reduction factor of 1.5 is applied to Kp (the coefficient of passive earth pressure) for cohesionless soils and to c (the cohesion term) for cohesive soils. Only stability conditions have been considered in the present study.

Anchored Sheet Piles in Cohesionless Soils

The pressure distribution diagram based on the conventional method in the absence of water in the backfill is shown in Figure 1. Following Bowles [16] the depth of penetration X of the sheet pile below the point of zero pressure is given by

$$2X^3 + 3X^2(h_3 + a) - \frac{6R_a\bar{y}}{G_k} = 0 \tag{1}$$

The force in the anchor rod P_{ar} is given by the following expression:

$$P_{ar} = R_a - R_p \tag{2}$$

When the anchored sheet piles are to be designed according to the BS8002 code [14] the pressure distribution diagram shown in Figure 2 is used. Stability of the wall is based on equilibrium, that is, the restoring moment should equal the disturbing moment. Also the mobilized soil strength, ϕ_{mob} is utilized and a surcharge of $10KN/m^2$ on the retained soil is applied. Furthermore a compulsory 0.5 m over-dig level or 10% of the total height retained is assumed, in order to provide for unforeseen and accidental events. From Figure 2, moment equilibrium is got by taking moments about the anchor rod level, yielding

$$R_a\bar{y} - R_p y' = 0 \tag{3}$$

where y 0.67D h ¢ = + , and D is the depth of embedment. For horizontal equilibrium an identical relationship to equation (2) is used.

Anchored Sheet Piles in Cohesive Soils

The analysis is based on the free-earth method using the conventional CECP2 methodology and using the BS8002 design procedure. For the current purpose it has been assumed that the backfill material is cohesionless and the sheet pile wall is only embedded in cohesive material. The pressure distribution diagram based on the conventional method for the anchored sheet pile in cohesive, or $\phi = 0$, material is shown in Figure 3. Following Bowles [16] the embedment depth D is obtained from the expression

$$D^2 + 2Dh_3 - \frac{2\bar{y}R_a}{4c - \bar{q}} = 0$$

Using the BS8002 design procedure the compulsory 10 KN/m2 surcharge is applied to the backfill up to the level of the dredge line. Also a fraction of the peak soil strength is used in the earth pressure calculations. In addition the unplanned excavation of not less than 0.5m or 10% of the retained height is incorporated.

RESULTS AND DISCUSSION

Results obtained for the analysis and design of anchored sheet piles in both cohesion less and cohesive soils are presented in Tables1 and 2 respectively. Every attempt has been made to ensure uniformity of design parameters in respect of the CECP2 and BS8002 methods in order to aid the comparative study. From Table 1 in relation to cohesion less soils, it is obvious that the embedment depths obtained using the conventional methods based on CECP2 and its variants are very similar although the factors of safety (F.S.) are very different. A 30% increase in the calculated embedment depth results in an F.S. of 2.04 as compared to an F.S. of 1.53 when a reduction factor is applied to the coefficient of passive resistance, Kp. In contrast the BS8002 design method gives a relatively slight increase in embedment depth. If the BS8002 method is adopted with an F.S. greater than unity, this gives rise to a greater embedment depth and a smaller force in the anchor rod. An increase in the embedment depth will result in an increase in the factor of safety and a decrease in the anchor rod force. However calculations not shown here suggest that for the present study, for an embedment depth in Table 1slightly in excess of 3.7m, the passive forces become greater than the active forces leading to negative values of anchor rod forces. This is impractical however and would suggest that the anchored sheet pile has an optimum embedment depth beyond which any further increase in depth becomes unreasonable. In Figure 4 the relationship between the factor of safety and the embedment depth based on the BS8002 method is presented. It is obvious that in the sheet pile under consideration, for embedment depths D below 3.5m, the variation between F.S. and D is reasonably linear. Beyond this point however relatively much larger increases in F.S. occur. As noted earlier for the present design study, an embedment depth D of 3.7m represents a practical maximum.

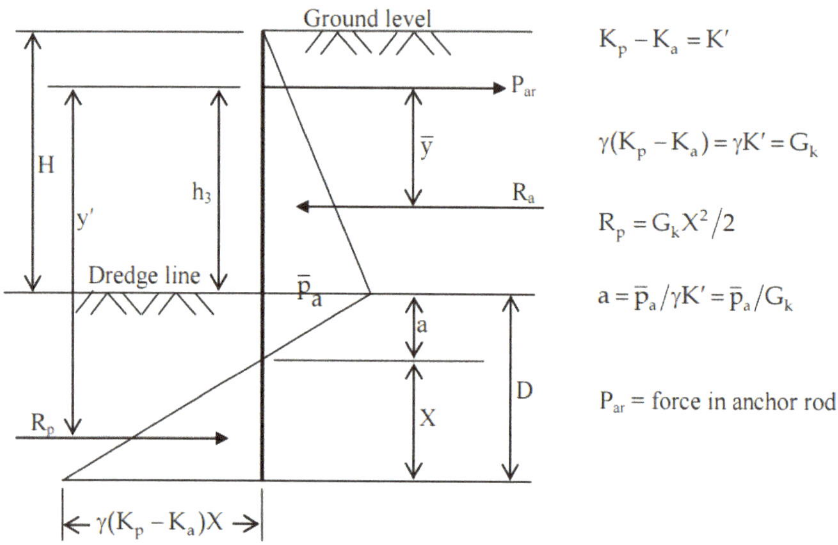

Figure 1: Pressure distribution diagram for anchored sheet pile in cohesionless soil, using the conventional method

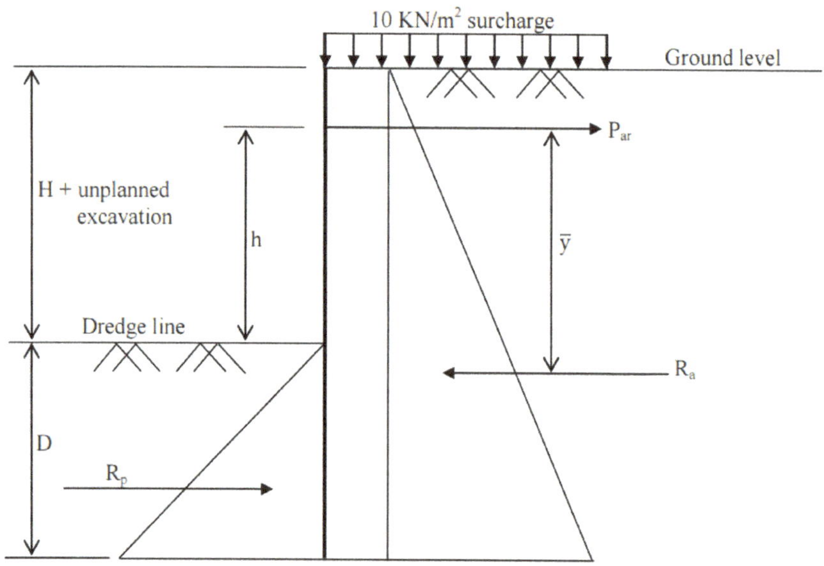

Figure 2: Pressure distribution diagram for anchored sheet pile in cohesionless soil, using the BS8002 method

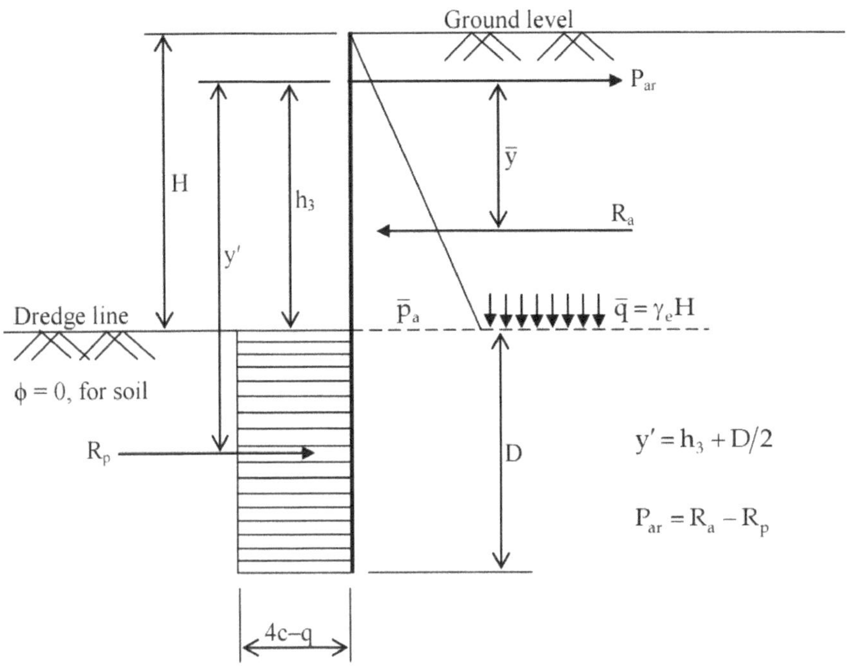

Figure 3: Pressure distribution diagram for an anchored sheet pile in cohesive soil

Table 1: Results for the design of anchored sheet pile in cohesionless soil

Method used	Design parameters	Embedment depth, D (m)	Factor of safety	Anchor rod force, P_{ar} (KN)	Total length of pile (m)
Conventional (CECP2) (a) Increasing computed depth by 30%	Retained height = 8.0 m. Anchor point is 1m below top of pile. γ =18 KN/m^3	2.52	2.04	30.3	10.52
(b) Reducing K_p by a factor (i.e. using $K_p/1.5$)	$\phi = 30^o$ $\delta = 20^o$	2.53	1.53	87.06	10.53
BS8002 approach based on taking different trial depths and subsequently checking equilibrium	Retained height = 8.7 m (i.e. 8.0 m +	2.8	1.00	141.43	11.5
	0.7m unplanned excavation).	3.0	1.10	113.07	11.7
	γ =18 KN/m^3 Design $\phi = 25.7^o$	3.5	1.30	57.28	12.2
	$\delta = 19.8^o$	3.7	1.46	7.57	12.4

Table 2: Results for the design of anchored sheet pile in cohesive soil

Method used	Design parameters	Embedment depth, D (m)	Factor of safety	Anchor rod force, P_{ar} (KN)	Total length of pile (m)
Conventional (CECP2) (a) Increasing computed depth by 40%	Retained height = 9.0 m. Anchor point is 1m below top of pile. For granular backfill, ϕ = 30° and γ = 16.5 KN/m^3. Below dredge line $\phi = 0$ and $c = 72.0$ KN/m^2	1.12	1.3	75.51	10.12
(b) Reducing the cohesion term by a factor (i.e. using c/1.5)		2.53	3.2	85.11	11.35
BS8002 approach based on taking different trial depths and subsequently checking equilibrium	Retained height = 9.8 m (i.e. 9.0 m + 0.8m unplanned excavation). For granular backfill, design ϕ = 25.7°, γ =16.5 KN/m^3. Below dredge line $\phi = 0$ and $c = 72.0$ KN/m^2	1.55	1.0	113.6	11.35
		1.70	1.11	96.2	11.5
		2.0	1.3	61.27	11.8
		2.3	1.5	26.38	12.1
		2.5	1.7	3.12	12.3

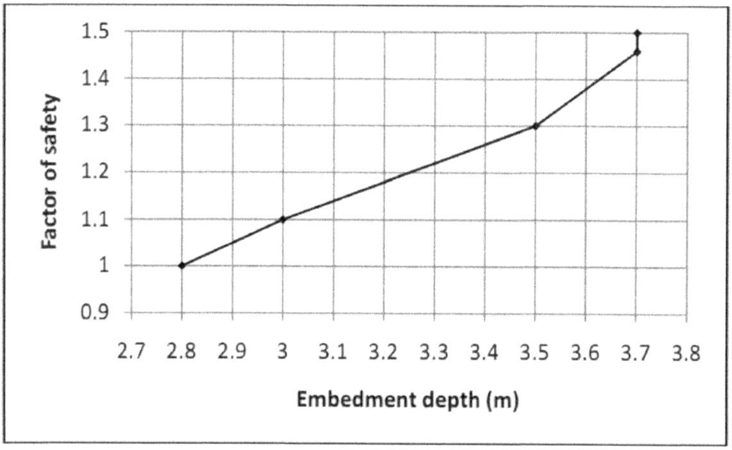

Figure 4: Variation of safety factor with embedment depth in granular soils - BS8002 method

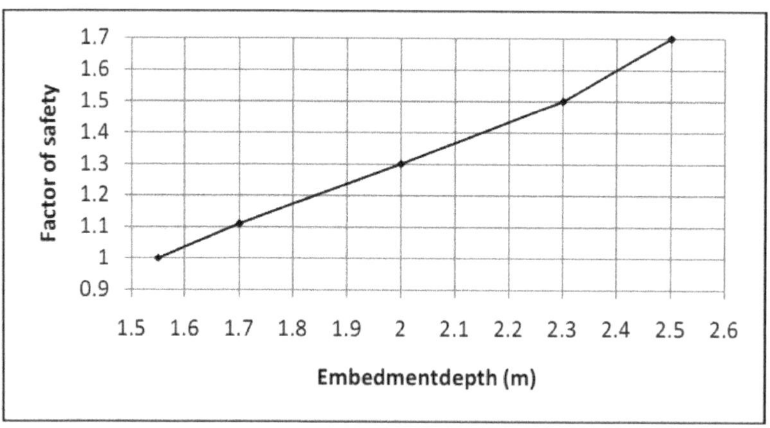

Figure 5: Variation of safety factor with embedment depth in cohesive soils - BS8002 method

In Table 2 dealing with embedment in cohesive soils the results show that the use of the conventional methods based on CECP2 and its variants gives greatly varying embedment depths depending on whether the factor of safety is applied to the computed depth or the cohesive term. Also the values of the F.S. obtained using the conventional methods differ greatly. On the other hand using the BS8002 approach assuming the worst credible loading results in an embedment depth approximately intermediate between the values given by the conventional methods, for the situation or condition when the factor of safety is unity. However for this equilibrium of moments condition, the force in the anchor rod is relatively large. From Table 2 it is apparent that using the BS8002 approach and increasing the embedment depth, the F.S. increases gradually while there is a corresponding decrease in the anchor rod force. It can be inferred that the wall becomes more stable with increase in the embedment depth, but this also leads to an increase in the overall length of the pile making it less economical. The graph of factor of safety against embedment depth in Figure 5 shows that the factor of safety increases almost linearly with increase in embedment depth for the BS8002 method. In summary from the results in Tables 1 and 2 it is apparent that for both granular and cohesive soils the BS8002 approach appears consistent since greater embedment depths lead to higher factors of safety but a corresponding decrease in the force of the anchor rod.

CONCLUSIONS

The present study sets out to investigate the applicability of the BS8002 limit state approach in addition to the conventional methods of CECP2 and its variants to the analysis and design of anchored sheet piles in granular and cohesive soils. From the results of the study the following conclusions have been drawn.

a) In the case of granular soils, use of the conventional procedures of either increasing the computed depth by 20% to 40% or applying a reduction factor to the coefficient of passive resistance will result in very similar embedment depths, but give rise to markedly different factors of safety. The BS8002 design approach in comparison produces only a relatively moderate increase in embedment depth, but lower factor of safety.

b) In the case of cohesive soils, the conventional methods give rise to greatly varying embedment depths depending on whether the factor of safety is applied to the computed depth or the cohesion term. In addition the calculated factors of safety using the conventional methods greatly differ. In contrast the BS8002 approach (for F.S. = 1.0) produces an embedment depth approximately intermediate between the estimates given by the conventional methods. Also for the equilibrium of moment's condition, the BS8002 value of anchor rod force is relatively large

c) The BS8002 design approach suggests that the factor of safety increases fairly linearly with increase in embedment depth for both granular and cohesive soils. However there is a corresponding decrease in the anchor rod force. The BS8002 method appears to be consistent in these respects. Notwithstanding, in the case of granular soils there appears to be an optimum embedment depth for anchored sheet piles beyond which any further depth increase becomes impractical.

REFERENCES

1. Murthy, V.N.S., 2002, Geotechnical Engineering: Principles and Practice of Soil Mechanics and Foundation Engineering, 1st Edition, Marcel Dekker Inc., New York, pp. 881–950.
2. Tschebotarioff, G.P., 1949, Large Scale Earth Pressure Tests with Model Flexible Bulkheads, Final Report to Bureau of Yards and Docks U.S. Navy, Princeton University, 112pp.
3. Civil Engineering Codes of Practice Joint Committee, 1951, Earth Retaining Structures: CECP2, Instn. Struct. Engs., London.
4. Rowe, P.W., 1952, Anchored Sheet Pile Walls, Proc. Instn. Civ. Engs., London, Part 1, Vol. 1, pp. 27–70.
5. Terzaghi, K., 1954, Anchored Bulkheads, Trans. Am. Soc. of Civ. Engs., Vol. 119, pp. 1243–1324.
6. Burland, J.B., Potts, D.M. and Walsh, N.M., 1981, The Overall Stability of Free and Propped Cantilever Retaining Walls, Ground Engineering, Vol. 14, No. 5, pp. 28–38.
7. Potts, D.M. and Burland, J.B., 1983, A Numerical Investigation of the Retaining Walls of the Bell Common Tunnel, Supplementary Report 783, Transport and Road Research Laboratory (TRRL), Crowthorne, England.
8. Padfield, C.J. and Mair, R.J., 1984, Design of Retaining Walls Embedded in Stiff Clay, Report 104, Construction Industry Research and Information Association (CIRIA), London.
9. Potts, D.M. and Day, R.A., 1990, Use of Sheet Pile Retaining Walls for Deep Excavations in Stiff Clay, Proc. Instn. Civ. Engs., London, Vol. 88, Part 1, pp. 899–927.
10. Clayton, C.R.I. and Bica, A.V.D., 1991, Discussion of Limit Equilibrium Methods for Free Embedded Cantilever Walls in Granular Materials by I.F. Symons, Proc. Instn. Civ. Engs., London, Vol. 90, Part 1, pp. 213–216.
11. Bolton, M.D., 1996, Geotechnical Design of Retaining Walls, The Structural Engineer, Instn. Struct. Engs. London, Vol. 74, No. 21, pp. 365–369.
12. Powrie, W., 1996, Limit Equilibrium Analysis of Embedded Retaining Walls, Geotechnique, Vol. 46, No. 4, pp. 709–723.
13. Craig, R.F., 2004, Craig's Soil Mechanics, 7 th Edition, Taylor and Francis, London, pp. 161–226.
14. British Standards Institution, 1994, Code of Practice for Earth Retaining Structures (BS8002), British Standards Institution, London.
15. Franklin, S.O. and Olopade, A.L., 2012, A Commentary on the Use of the CECP2 and BS8002 Methods for the Analysis and Design of Cantilever Sheet Piles,

Journal of Basic and Applied Scientific Research, TextRoad Publication, Vol. 2, No. 2, pp. 1222–1229.

16. Bowles, J.E., 1988, Foundation Analysis and Design, 4th Edition, McGraw-Hill, New York, pp. 580–643.

16

CHAPTER

CONTIGUOUS PILE WALL AS A DEEP EXCAVATION SUPPORTING SYSTEM

Venkata Ramasubbarao Godavarthi, Dineshbabu Mallavalli, Ramya Peddi, Neelesh Katragadda, and Prudhvikrishna Mulpuru

Keywords: Deep Excavation, Excavation supporting system, Contiguous Pile wall.

INTRODUCTION

Urbanization made cost of land high and it necessitated to go deeper into the ground and also towering vertically towards the sky. A deep excavation into the ground is indispensible to create additional floor space to meet increasing space requirements for parking for multi-storey buildings at the town centres. Numbers of deep excavation pits in city centers are increasing every year [1]. Structures in the immediate vicinity of excavations, dense traffic scenario, presence of underground obstructions and utilities have made excavations a difficult task to execute [2]. In this context, analysis and design of proper deep excavations and their supporting systems are essential. Even in complicated urban settings, deep retaining systems have been deployed successfully by overcoming construction challenges. An Excavation is basic phase in the construction of foundations or basements of high rise buildings, underground oil tanks, subway stations etc [3]. The process of an excavation may encounter different kinds of soils underneath the same excavation site-from soft clay to hard rocks. During excavation, some soil types pose greater problems than others. Sandy soil is always considered dangerous even when it is allowed to stand for a period of time after a vertical cut. Vibration from blasting, traffic and heavy machinery movement, and material loads near the cut can also cause earth to collapse in sandy soil. The instability can be caused by moisture changes in the surrounding air or changes in the water table. Clayey soils in

Citation: Venkata Ramasubbarao Godavarthi, Dineshbabu Mallavalli, Ramya Peddi, Neelesh Katragadda, And Prudhvikrishna Mulpuru, Contiguous Pile Wall As A Deep Excavation Supporting Systemm, Issue 19, July-December 2011 p. 144-160, ISSN 1583-1078.

general, present less risk than sand; however, soft clay can prove to be very treacherous. Silty soils are also unreliable and require the same precautions and support provision as sand. To engineer an excavation, the basic steps which should be carried out by design engineer are: site characterisation; selecting dimensions of excavation; surveying adjacent structures; establishing permissible movements; selecting earth retaining system; selecting supporting and construction scheme; predicting movements; compare predicted with permissible movements; alter supporting (bracing) and construction scheme; if needed; monitor instrumentation, alter as needed; compare monitored results with predicted and permissible values; alter bracing and construction scheme, if needed [4, 5]. Site characterisation is the first major step to be taken by a geotechnical engineer. Most common practices tend to greatly simplify soil profile and to select appropriate design parameters like strength, and compressibility as the basis of simple laboratory tests. Since deep excavation is a total technique, proper coordination and integration of design and construction are of utmost important [3]. In this study, details of various deep excavation supporting systems are furnished and comprehensive design philosophy of an excavation supporting systems is presented. This paper also contains details of a case study on the analysis and design of a Contiguous pile wall for supporting deep excavation at a town centre.

MATERIAL AND METHOD

Deep Excavation Supporting Systems

Unsupported excavations pose hazard to workers and equipment. Prevention and minimising damages to surrounding is of utmost concern to the design engineers and constructor for any excavation work. A variety of excavation methods and lateral supporting systems are to be practiced based on local soil, ground water and environmental conditions, allowable construction period, money and machinery. Excavation methods include full open cut methods, braced excavation methods, anchored excavation methods, island excavation (partial excavation) methods, and top-down construction methods and zoned excavation methods. Types of deep vertical soil support systems [3, 6] are commonly used in metropolitan cities are Conventional retaining walls (figure 1), Soldier pile with wooden lagging walls (figure 2), Sheet pile walls (figure 3, 4 and 5), Diaphragm walls (figure 6) and Pile walls-Contiguous, Secant or Tangent (figure 7). Apart from retaining walls to resist lateral earth pressure a supplementary strutting systems are also required. A strut is made of wood, reinforced concrete or steel. Based on function of a strut, it may classify as an earth berm, a horizontal strut, an anchor or as a top-down floor slab. During construction of excavation supporting system, the adjacent facilities may damage. Vibration due to adjacent machinery, vehicles, rail-roads, blasting and other sources require that additional bracing precautions are to be taken. This may avoided by taking some ground improvement measures such as grouting the ground between the excavation site and adjacent building. Dewatering should be done if the soil at the site is relatively impervious like clayey soil.

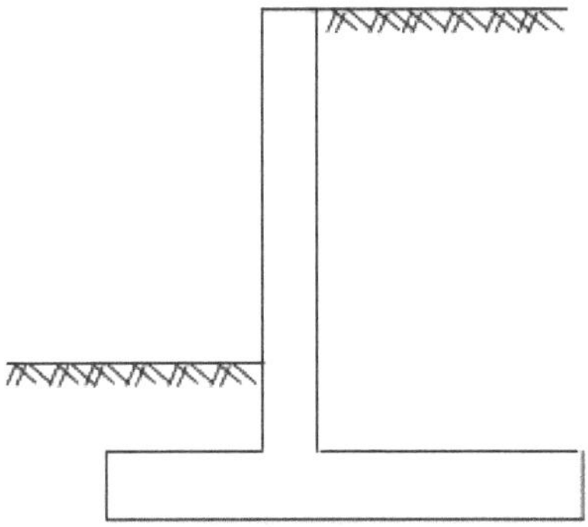

Figure 1: Retaining wall

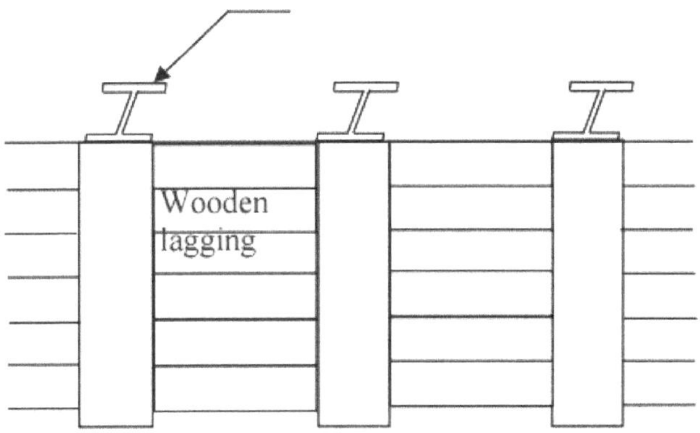

Figure 2: Soldier pile with wooden lagging walls

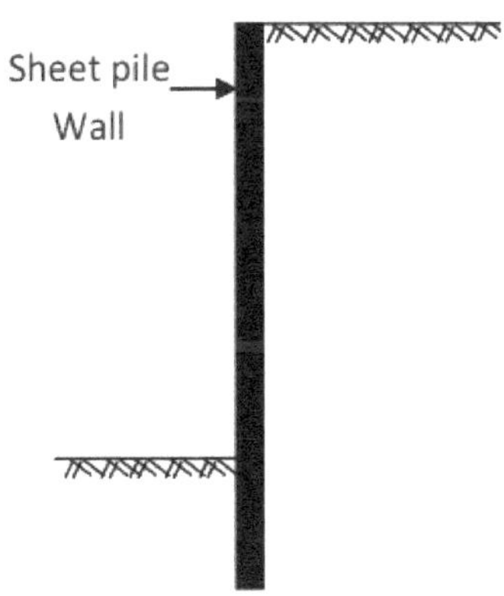

Figure 3: Sheet pile wall

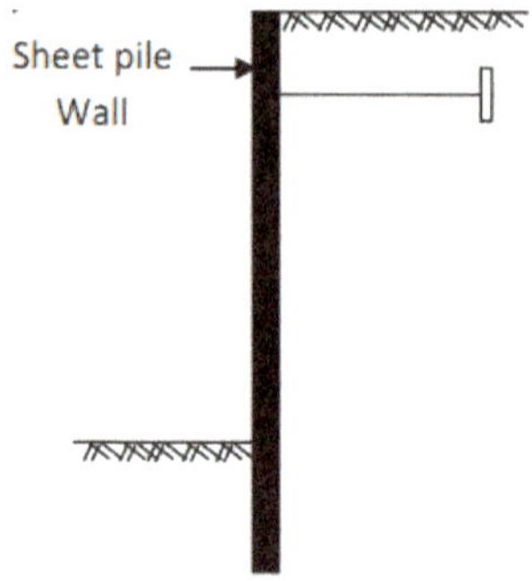

Figure 4: Anchored Sheet pile wall

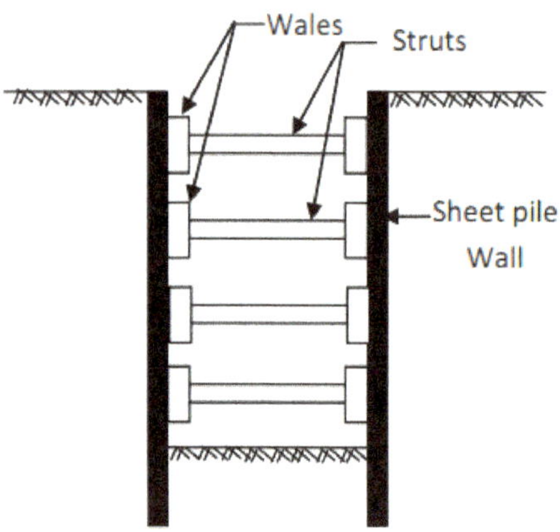

Figure 5: Braced wall

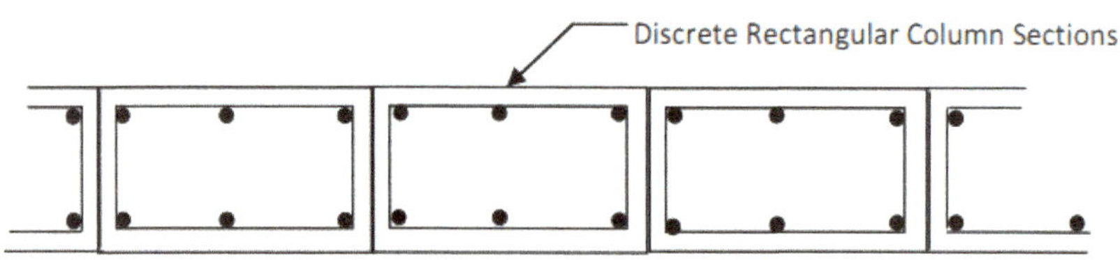

Figure 6: Diaphragm Wall

Embedded walls take up little lateral space as compared to the conventional retaining walls. Soldier pile and lagging walls are the most economical systems compared to other retaining walls. In granular soils where the groundwater is below proposed subgrade or can be economically drawn down by dewatering, a soldier beam and lagging system remains the most widely used approach. This type of retaining system involves the following broad based activities: (i) Constructing soldier piles at regular intervals (1 to 3m on centre typically) (ii) Excavating in small stages and installing wooden lagging. (iii) Backfilling and compacting void space behind the lagging. Moment resistance in soldier pile and lagging walls is provided solely by the soldier piles. Passive soil resistance is obtained by embedding the soldier piles beneath the excavation grade. The lagging bridges and retains soil across piles and transfers the lateral load to the soldier pile system. They are also very easy and fast to construct. The major disadvantages of soldier pile and lagging systems are that they are primarily limited to temporary construction. They cannot be used in high water table conditions without extensive dewatering. They are not as rigid as other retaining systems because flange of the soldier pile is only embedded beneath subgrade. When soils are not conducive to dewatering, sheet piling capable of withstanding the hydrostatic pressure may be the more appropriate option. For deep cuts, lateral restraint can be provided by horizontal wales with either raker braces or by tieback anchors. For shallow excavations, a cantilevered wall is generally sufficient to resist earth pressures. Narrower excavations can be braced with cross struts. Diaphragm walling is a technique of constructing a continuous underground wall from the ground level. These reinforced concrete diaphragm walls are also called Slurry trench walls due to the reference given to the construction technique where excavation is made possible by filling and keeping the wall cavity full with bentonite-water mixture during excavation to prevent collapse of vertical excavated surfaces. Typical wall thickness varies between 0.6 to 1.1m. The wall is constructed panel by panel in full depth. Panel width varies from 2.5m to about 6m. Short widths of 2.5m are selected in less stable soils, under very high surcharge or for very deep walls. It must be remembered that Diaphragm walls are constructed as a series of alternating primary and secondary panels. Alternate primary panels are constructed first which are restrained on either side by stop-end pipes. Before the intermediate secondary panel excavation is taken up, the pipes are removed and the panel is cast against two primary panels on either side to maintain continuity. The major disadvantage of Diaphragm wall is it requires massive equipment, long construction period, and huge cost.

Piled Retaining Walls

In-situ pile retaining walls also called column piles are rows of concrete piles either cast-in situ pile method or precast pile method. Merits of column piles are less noise or vibration than produced by the installation of solider piles or sheet piles. Colum piles have greater stiffness than soldier piles or steel sheet piles. They avoid excessive bulk excavation and help to control ground movements. There are three distinct bored pile wall options in current use: Contiguous wall, Secant wall and tangent wall. Contiguous pile walls are constructed with small gaps between adjacent piles. The use of low-cost augers and, more particularly, Contiguous Flight Auger (CFA) rigs to drill successive unconnected piles provide an economical wall. Diameter and spacing of the piles is decided based on soil type, ground water level and magnitude of design pressures. Large spacing is avoided as it can result in caving of soil through gaps. CFA pile diameters range from 300mm to 1000mm. CFA piles are considered more economical than diaphragm wall in small to medium scale excavations due to reduction in cost and time of site operations. Besides, no bentonite mud is needed for the excavation. Contiguous piles are suitable in crowded urban areas, where traditional retaining methods would otherwise encroach the adjoining properties, these piles restricts ground movements on the backfill side. The pile is formed by first drilling into the ground with a CFA. Cement-sand grout or concrete is then injected under pressure through the auger's hollow stem as it is being withdrawn. The grout or concrete pressure is maintained during the auger withdrawal so that it assists the extraction as well as exerting a lateral pressure on the surrounding soils. On completion of this operation, a reinforcing cage is placed into the fluid column of grout or concrete. When CFA pile combined with capping beams/ breasting beams can show savings in cost and time. Capping beams at the top to help equitable pressure distributions in piles. Separate facing usually provided to improve looks. The range of soil conditions in which CFA piles can be used are granular soils, cohesive soils, soft rocks. Soft clays, weak organic soils are unsuitable due to wall bulging. Hard rocks are also not suitable. The Contiguous wall can only be used where ground water is not a hazard or where grouting or jet grouting is used can be used to remedy leakage between the piles. However, some acceptable amount of water can be collected at the base and pumped out. The principal disadvantages of contiguous pile walls-the gaps between piles and the resulting problems of lack of water proofness have been effectively overcome by interlocking or secant piles.

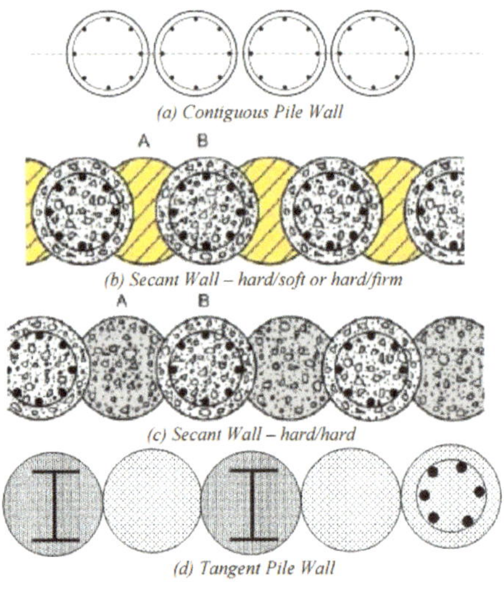

Figure 7: Pile Walls

Secant Pile Walls are formed by constructing intersecting piles. Secant bored pile walls are formed by keeping spacing of piles less than diameter. Secant pile walls are used to build cut off walls for the control of groundwater inflow and to minimize movement in weak and wet soils. Secant Wall constructed in the form of hard/soft or hard/firm and Secant Wall Hard/hard wall. Secant Wall-hard/soft or hard/firm is similar to the contiguous bored pile wall but the gap between piles is filled with an unreinforced cement/bentonite mix for the hard/soft wall and weak concrete for the hard/firm wall. Construction is carried out by installing the primary piles (A) and then the secondary piles (B) are formed in reinforced concrete, cutting into the primary piles. Diameters can range from 500mm to 1200mm. Secant Wall Hard/hard wall construction procedure is very similar to a hard/firm wall but in this case the primary piles (A) are constructed in high strength concrete and may be reinforced. The Secondary piles (B) are cut into the concrete primary piles (A) using heavy duty piling rigs fitted with specially designed cutting heads. Tangent pile walls consist of a series of drilled shafts located such that the adjacent shafts touch each other, hence the name tangent wall. Secant pile walls are stiffer than tangent piles walls and are more effective in keeping ground water out of the excavation.

RESULTS AND DISCUSSION

Design Methodology Involving Excavation Retaining Systems

Stability of excavation is the major design criterion in order to avoid collapse of excavations [3]. Stability analysis involves the distribution of earth pressures. Stability analyses include push-in failure analysis, sand boiling analysis, and upheaval analysis. To excavate in a sandy soil one should consider push-in and sand boiling. To excavate in a clayey soil one should consider push-in and upheaval. In alternated layers of sand (or gravel) and clay, push-in and upheaval should consider. The penetration depth of a retaining wall is usually determined according to results of push-in failure because sand boiling is normally not a main controlling factor. Goals of lowering the ground water level are to keep the excavation bottom dry, to prevent leakage of water or sand, to avoid sand boiling or upheaval failure, and to prevent the occurrence of floating basements. Distribution of earth pressure influences stability, stress, and deformation analysis of the deep excavation. For problems of excavation, considering that the active earth pressure is usually the main force leading to the failure of the excavation supporting systems. The passive earth pressure is usually the force resisting the failure. The pressure distribution shall depend on the nature of backfill. An excavation will encounter various soil layers. Earth pressure computed on the basis of Rankine's earth pressure theory. If working with long term behaviour, effective stress analysis should be applied to both cohesive and cohesionless soils and water pressure should be computed separately. Recharge wells may be required outside the walls if drops in water level are not allowed. If there are compressible soils in the profile, buildings and lifeline structures are expected to settle due to groundwater lowering. Dewatering methods are applied in cases where there is need for control of water pressures and/or flow [7]. There are two analysis methods for push-in failure: free-earth support method and fixed earth support method [8, 9]. The free earth support method assumes that the embedment of retaining wall is allowed to move a certain distance under the action of lateral earth pressure. The fixed earth support method is to assume that embedment of the retaining wall seems to be fixed at a point below the excavation surface. The embedded part may rotate about fixed point. Thus, when the retaining wall is in the limiting state, the lateral earth pressure around the fixed point on the two sides of the retaining wall does not necessarily reach the active or passive pressures. If a cantilever wall is designed based on the free support method, no fixed point is supposed to exist in the embedded part of the wall, as discussed above. The external forces, only active and passive forces, on the retaining wall are not to come to equilibrium. Therefore, the free earth support method is not applicable to cantilever walls. A simplified analysis [10] of considering active force on one side and passive force on the other side of excavation supporting system is used to simplify the calculation of necessary depth of penetration or depth of embedment (DP). DP is found by applying moment equilibrium about the bottom of the wall. The computed may be increased by 20 to 40% beyond the point required by equilibrium or the effective horizontal pressure on the passive side may be reduced by applying a factor of safety of 1.5 to 2.0 before the embedment depth of pile is computed. The upward flow may cause the effective stress at a point in soil to be zero, which means that the soil is unable to bear any load, and the phenomenon is called sand boiling. The hydraulic gradient when the effective stress equals zero is called critical hydraulic gradient (i_{cr}), which can be expressed as follows:

$$i_{cr} = \frac{\gamma^1}{\gamma_w} \tag{1}$$

The factor of safety against boiling can be estimated by Harza's method, Terzaghi's method or simplified one-dimensional method. For excavations, a reasonable factor of safety is around 1.5-2.0 [11]. The simplified one-dimensional method is presented in figure 8.

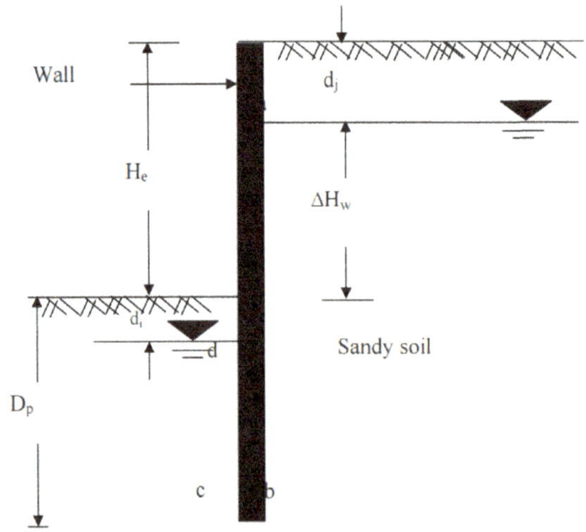

Figure 8: Analysis of Sand boiling

$$i_{avg} = \frac{\nabla H_w}{H_e - d_j + d_i + 2(D_p - d_i)} = \frac{\nabla H_w}{(H_e + 2D_p - d_i - d_j)} \tag{2}$$

The factor of safety against sand boiling will be:

$$F_s = \frac{i_{cr}}{i_{avg}} = \frac{\gamma^1(H_e + 2D_p - d_i - d_j)}{\gamma_w \nabla H_w} \tag{3}$$

The required Fs for the above equation should be greater than or equal to 1.5. If below the excavation surface there exists a permeable layer (such as or gravel soils) underlying an impermeable layer, the impermeable layer has a tendency to be lifted by the water pressure from the permeable layer. The safety, against upheaval, of the impermeable layer should be examined. As shown in figure 9, the factor of safety against upheaval is:

$$F_{up} = \frac{\Sigma_i \gamma_{ti} h_i}{H_w \cdot \gamma_w} \tag{4}$$

where, F_{up}=factor of safety against upheaval, γ_{ti}= unit weight of soil in each layer above the bottom of the impermeable layer, h_i = thickness of each soil layer above the bottom of the impermeable layer, H_w= head of the water pressure in the permeable layer, and γ_w = unit weight of the ground water. The factor of safety against upheaval F_{up} should be larger than or equal to 1.2. The factor of safety against of excavation construction, the possibilities (of the occurrence) of upheaval at each stage of excavation should be analyzed. If drilling within the excavation zone is required (e.g. in order to place piezometers or build a well), the possible paths of water flow should be verified and the possible upheaval induced by drilling should be prevented to secure the excavation. With the stability analysis of an excavation, the penetration depth of retaining wall can be determined. With the stress and deformation analysis, the bending moment and shear of the retaining wall can be determined. The thus obtained bending moment and shear envelopes are then used for the design of reinforced concrete columns. Basically, the stiffness of the reinforced concrete pile is the highest. Colum piles bear the axial load and flexural load simultaneously. Therefore, their behavior is similar to that of the reinforced concrete columns.

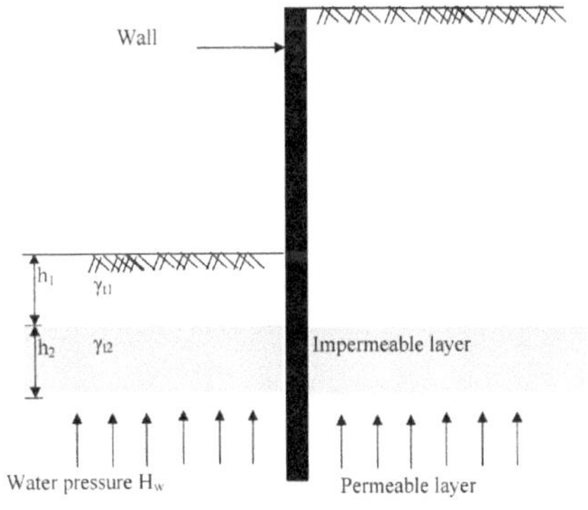

Figure 9: Analysis of Upheaval

Contiguous Pile Wall as Deep Excavation Supporting System at a Commercial Complex in Vijayawada, Andhra Pradesh, India – Case History A supporting system for proposed 11.2m deep excavation was necessary for a commercial complex at Vijayawada, India. The planning of sub-soil has been carried as per the Indian standard codal provisions [12]. Five bore holes of 25m deep are taken in such a way that one at centre and one at each corner of the site. Average profile and properties of soil layers are evaluated as per Indian standard soil laboratory testing and classification procedures [13, 14] and are presented in figure 10. Presence of footings of adjacent buildings and underground utilities were some of the decisive factors that eliminated choice of diaphragm wall for supporting the excavation. In this situation, contiguous pile wall for supporting the excavation is the right the choice.

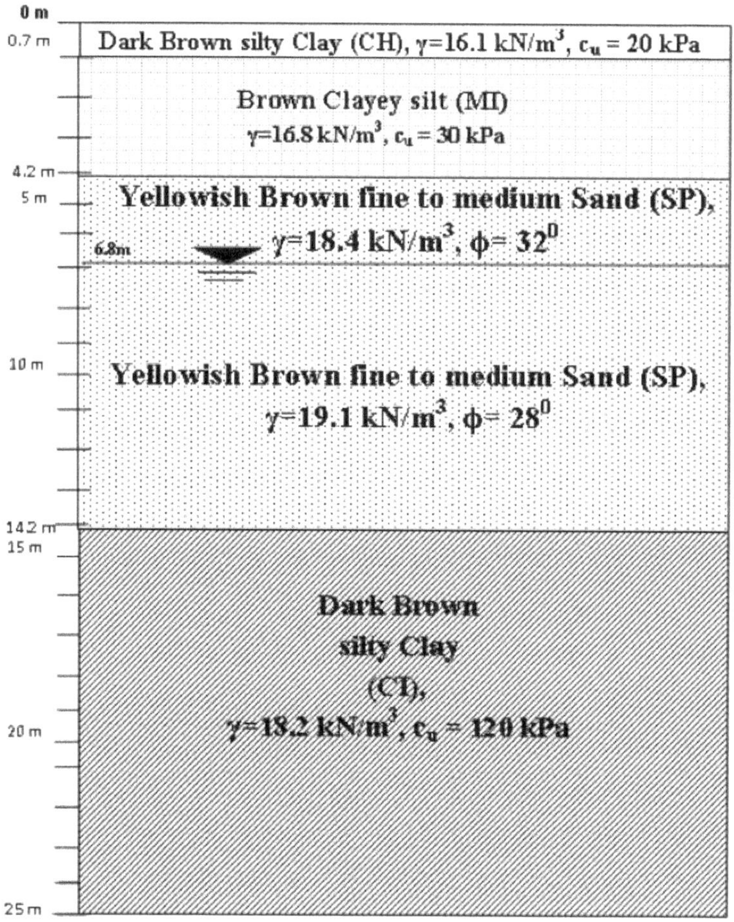

Figure 10: Bore Hole

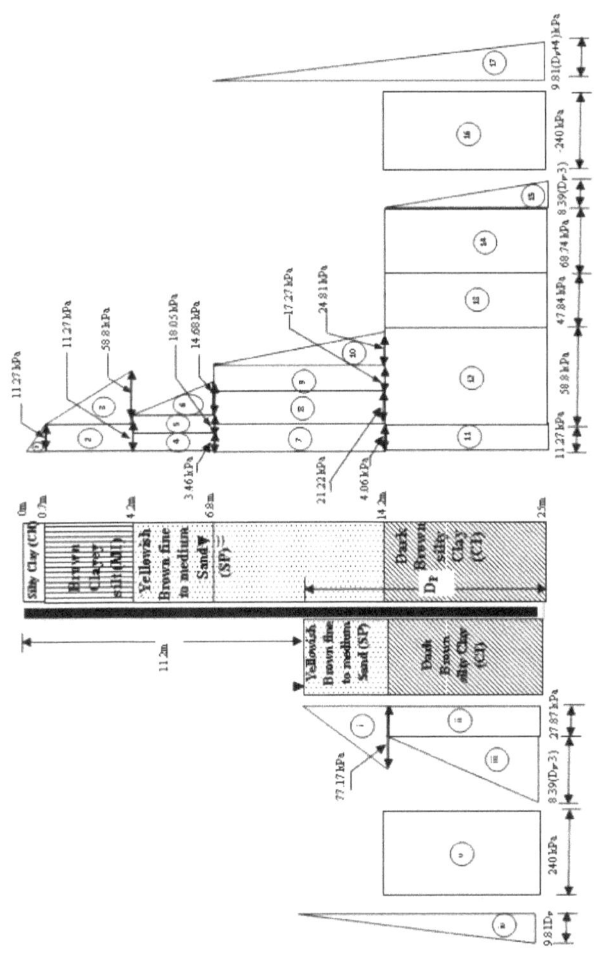

Figure 11: Lateral Earth Pressures acting on the Wall

Detailed lateral earth pressures acting on the wall are calculated and shown in figure 11. Stability analysis of the wall has been carried to evaluate the depth of penetration (DP) by equating active pressure on right side of the wall to passive pressure on left side of the wall. For the present case, DP is 5.64m. By applying factor of safety of 1.5 to coefficient of passive earth pressure, DP is 7.79m. An actual penetration depth (DP) of 9.8m (=1.25×7.79m) is adopted. The factor of safety against sand boiling is 5.16 obtained by using equation (3) with consideration of sand layer throughout the penetration depth. This value is greater than 1.5. Hence, wall is stable against sand boiling also. The diameter of pile used is 600mm. M30 grade concrete and Fe500 grade steel are used for making each pile. For contiguous piles, the c/c spacing is used for estimating reinforcement quantity. The pile is designed to resist the moment caused by the lateral earth pressure. The pile is designed to resist a maximum bending moment of

726.82 kN-m. This moment is evaluated from deducting moment due to the active force and moment due to the passive force with respect to pile bottom. Area of reinforcement required for the pile section is 10,179 mm2 which can be obtained from Indian standard design aids [15]. A clear cover of 50mm provided to the pile section. The required area of main reinforcement is satisfied by providing 14 bars of 32mm diameter. The strength of the core concrete is enhanced by the confinement through spiral (helical) reinforcement [16, 17]. Therefore, helical reinforcement of 8mm Diameter and 50mm Pitch are provided. Reinforcement details are presented in the figure 12.

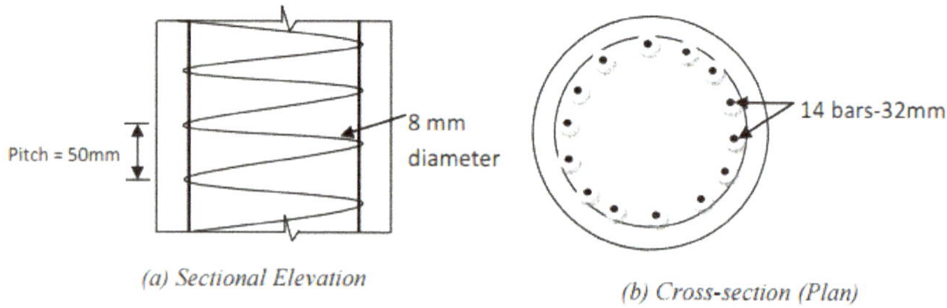

Figure 12: Reinforcement Details

The pile wall comprised 600mm diameter bored cast-in-situ piles spaced at 0.7m c/c, with an average embedment depth of 9.8m. These shall be installed vertically along the proposed line of excavation. These piles were designed as cantilever retaining system without provision of anchors. Capping beam of 1200 x 750mm size was cast along the alignment of the pile which was designed to take care of differential changes in the earth pressures in the adjacent piles. The peripheral length of excavation was 310m and 442 piles were involved in the support work. Individual piles were constructed in the same way as a typical Bored Castin-situ piles using temporary casings and bentonite slurry. Excavation operations proceeded only in those areas where pile concrete attained maturity. General construction sequence for the piling operation is: (i) Centering of rotary rig on the proposed pile point (ii) Carrying out the boring operation upto the upto about 21m below the existing ground level (iii) Driving the casing if needed (iv) Maintaining the stability of the borehole simultaneously with bentonite slurry (v) Continuing boring operation in soil using soil bucket and/or Soil auger depending on the stratum (vi) After completion of boring, cleaning of borehole by bentonite flushing (vii) Lowering of reinforcement cage into the borehole (viii) Repeat bentonite flushing operation and subsequently (ix) Pouring M30 grade concrete through tremie. The water drained out shall be let away from the site such that it does not re-enter in to the site. If ground water table is encountered in the region of operation, then suitable well point dewatering system should be activated to avoid flooding and subsequent collapse. Plate 1 shows the contiguous pile wall constructed at the site.

Plate 1: Contiguous Pile Retaining System for Commercial a Complex Vijayawada

CONCLUSIONS

Lack of inadequate space at town centres goes for deep vertical excavations, which require supports that are designed to consume minimum construction space. Various excavation supporting systems are presented in this study. Contiguous pile wall is one good choice of earth retention systems in the urban context. The detailed analysis and design of a contiguous pile wall is explained by a Case study which uses contiguous pile wall as an excavation supporting system at an urban centre.

ACKNOWLEDGEMENTS

Authors are thankful to Dr. P.V. Narasaiah, Director of SRK Institute of Technology, Enikepadu, Vijayawada, Andhra Pradesh, India for his steady guidance and support.

REFERENCES

1. Ergun M.U., Deep Excavations, Electronic Journal of Geotechnical Engineering, Available at: www.ejge.com/Bouquet08/UfukErgun_ppr.pdf, 2008.
2. Manish Kumar, Deep support systems using diaphragm walls and contiguous piles, National Seminar on Deep Excavation in Urban Environment, Mumbai, 2008, p. 1-18.
3. Chang-Yu Ou, Deep Excavation: Theory and Practice, Taylor and Francis Group, London, UK, 2006.
4. Puller, M., Deep Excavations: A practical Manual, Thomas Telford, London, UK, 1998.
5. Moh Z.C., Chin C.T., Deep Excavation in Soft ground, Proceedings of ASIA Construction: New Frontiers, Singapore, 1991.

6. BD 42/00, Design of Embedded Retaining Walls and Bridge Abutments, Department for Regional Development, Scottish Executive Development Department, Northern Ireland.
7. Bureau of Indian Standards, Guidelines for Dewatering during Construction (IS: 9759), New Delhi, India, 1981.
8. Das B.M., Principles of Geotechnical Engineering, Cengage Learning, CT, USA, 2010.
9. Clayton C.R.I., Milititsky J., Woods R.I., Earth Pressure and Earth Retaining Structures, Blackie Academic & Professional, London, 1993.
10. Teng W.C., Foundation Design, Prentice Hall, Englewood Cliffs, NJ, 1962.
11. NAVFAC, Foundations and Earth Structures, Design Manual 7.02, Department of the Navy Naval Facilities Engineering Command, Alexandria, Virginia, 1982.
12. Bureau of Indian Standards, Code of practice for Subsurface Investigations for foundations (IS: 1892), New Delhi, India, 1979.
13. Bureau of Indian Standards, Compendium of Indian Standards of Soil Engineering, Laboratory Testing of soils for civil engineering (SP 36 Part 1), New Delhi, India.
14. Bureau of Indian Standards, Classification and Identification of soils for general engineering purpose, IS 1498, New Delhi, India, 1970.
15. Bureau of Indian Standards, Design Aids for Reinforced concrete to IS: 456-1978 (SP 16), New Delhi, India.
16. Dayaratnam P., Limit State Design of Reinforced Concrete structures, Oxford & IBH Publishing Co. Pvt. Ltd., New Delhi, India, 2004.
17. Shah H.J., Reinforced Concrete Vol.1 (Elementary Reinforced Concrete), Charotar Book House, Anand, Gujarat, India, 2005.

17
CHAPTER

INCREMENTAL DYNAMIC ANALYSIS OF SHEET PILE QUAY-WALLS IN ORDER TO REACH PERFORMANCE-BASED DESIGN CRITERIA FOR THIS TYPE OF WHARF

Reza Dezvareh, Prof. Khosrow Bargi and Yasin Moradi

Keywords: Sheet pile quay walls, incremental dynamic analysis, Soil-Structure interaction, IDA curves

INTRODUCTION

Quays are of essential parts of each port. Among all types, Sheet Pile Quay-walls are of major importance. Along with the development of marine transportations through Iranian long sea borders in north and south of the country, usage of modern design techniques are inevitable. Due to the high seismic vulnerability of the country, this paper tries to present author's new dynamic design achievements for Sheet Pile Quay-walls. Recent marine structure design codes i.e. Pianc [1] and OCDI [2] suggest performance based design (PBD) of various marine structures including quays. As well, current study concerns modern dynamic analysis technique of Sheet Pile Quay-walls, hoping for its contribution to assessment of quay performance criteria. Incremental Dynamic Analysis (IDA) is an advanced nonlinear dynamic analysis method that has absorbed much attention in the design of steel frames. It involves monitoring structural response from elastic to limit state through scaling ground motions. One of the most important applications of this method is assessing structure's behavior and design criteria in Performance-Based Earthquake Engineering (PBEE). Current study began with modeling of a conventional sheet pile quay-wall followed by dynamic analysis concerning soil-structure interaction and assessing the IDA curves and summarizing the results and in the end, Mean annual frequency (MAF) investigated.

Citation: Reza Dezvareh, Prof. Khosrow Bargi & Yasin Moradi, Incremental Dynamic Analysis Of Sheet Pile Quay-Walls In Order To Reach Performance-Based Design Criteria For This Type Of Wharf, http://www.arpapress.com/Volumes/Vol10Issue3/IJRRAS_10_3_09.pdf.

ABOUT INCREMENTAL DYNAMIC ANALYSIS

In this method, the model is exposed to one or several ground motions of different levels for each; structure's response curve is plotted. IDA curve of a structure is a curve for which an Intensity Measure (IM) like platform drift or maximum understory drift ratio is plotted versus damage measure (DM) or peak ground acceleration (PGA) or spectrum acceleration of first mode with 5% of damping. Limit-states and probability of exceeding a specified state can then be assessed through summarization of the obtained curves.

SEISMIC DESIGN FEATURES OF MARINE STRUCTURES CODE

Overseas Coastal Area Development Institute of Japan (OCDI) [2] is one of the most conventional codes for the design of Sheet Pile Quay-walls. It suggest to main earthquake levels for the design of quays:

EQ Hazard Level 1

Occurrence probability in service life of structure is about 20 to 50 percent (equal to an earthquake with the return period of 75 years), during which all modules of the port will continue their performance in elastic state.

EQ Hazard Level 2

Occurrence probability in service life of structure is about 10 to 20 percent (equal to an earthquake with the return period of hundreds of years), during which all modules of the port shall meet sufficient energy absorption in massive shakings.

PROBLEM DESCRIPTION AND MODELING

Sheet Pile Quay-wall

The length of the studied quay was 20.0 meter while its penetration depth was 10.0 meters. As installation of joint vertical piles is very formal, they were accordingly modeled in this study.

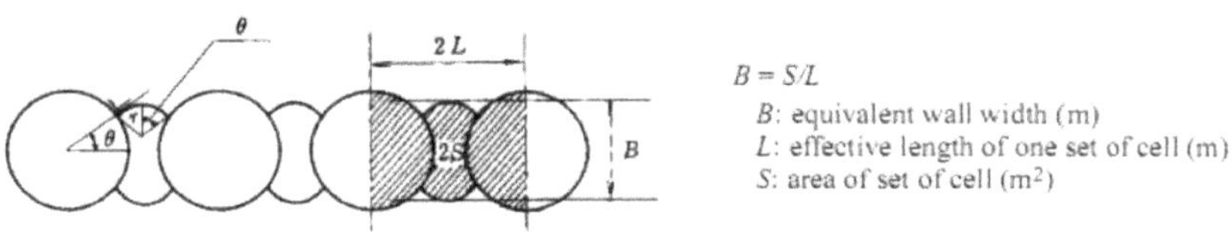

Figure 1: Section of sheet pile quay-wall

Soil Model and its Interaction with the Quay

Similar to other footing types, soil-structure interaction plays an important role in the analysis of marine structures. This role gets especially significant when concerning uplift capacity, overturning and slide of the structure. Nevertheless, soil body is the source of different forces imposed to the structure. For this part of finite element analysis (FEM), soil was modeled using solid elements. Moreover, frictional properties and internal constrains of different parts were applied. Friction between soil and structures' elements were modeled by Penalty method.[3]

Material Properties

Material properties include concrete, steel and soil which were applied to platform, steel quay and soil, consequently.

- Concrete properties: Elasticity modulus of 2.08e7 kPa, Poison's ratio of 0.3 and unit weight of 2400 kg/m3
- Steel properties (St-52):Elasticity modulus of 2.07e8 kPa, Poison's ratio of 0.3 and unit weight of 7850 kg/m3
- Soil properties:

Table 1: Soil properties

	unit weight	Yung's modulus	Pisson's ratio	Permeability	Cohesion	Internal friction angle
	kg/m3	E_0	ν	k	C	φ
Sand	19.5	25	0.25	0.8e-5	0	36
Clay	19.5	25	0.25	0.8e-6	25	0

Introduction Finite Element Software and Loading

Capable of modeling various soil and structural elements and interactions, ABAQUS [4] has been used for the finite element dynamic analysis in this study. According to the table, ten different records were used for dynamic analysis, each of which were first scaled in NERA [5] and then applied to bed rock. The resultant soil response was finally applied to corresponding nodes.

Table 2: Properties of 10 used records

No	Event	Station	PGA (g)
1	Imperial Valley, 1979	Plaster City	0.042
2	Loma Prieta, 1989	WAHO	0.37
3	Superstition Hills, 1987	Wildlife Array	0.18
4	Imperial Valley, 1979	El Centro Array	0.117
5	Loma Prieta, 1989	Coyote Lake Dam	0.179
6	Superstition Hills, 1987	Wildlife Array	0.2
7	Imperial Valley, 1979	Westmoreland Fire Station	0.11
8	Loma Prieta, 1989	Sunnyvale Colton Ave	0.207
9	Superstition Hills, 1987	Hollister Diff. Array	0.269
10	Imperial Valley, 1979	Chihuahua	0.254

Boundary Conditions

Reflecting the energy waves by boundaries makes geotechnical models far from reality. To meet this phenomenon, absorbing boundaries were introduced to the model. The so called absorbing boundaries have been defined by Lymser&Kuhlmeyer (1969). Damping coefficient of elements *Cd* used in this analysis was calculated as:

$$C_d = \rho V_s A \tag{1}$$

- ρ: special gravity of soil
- V_s: Shear wave velocity
- *A*: Covered area of each element

RESULTS OF INCREMENTAL DYNAMIC ANALYSIS (IDA)

Following figures present the results of incremental dynamic analysis for two types of constrained (with lateral support) and cantilever (with no lateral support) sheet pile quay wall.

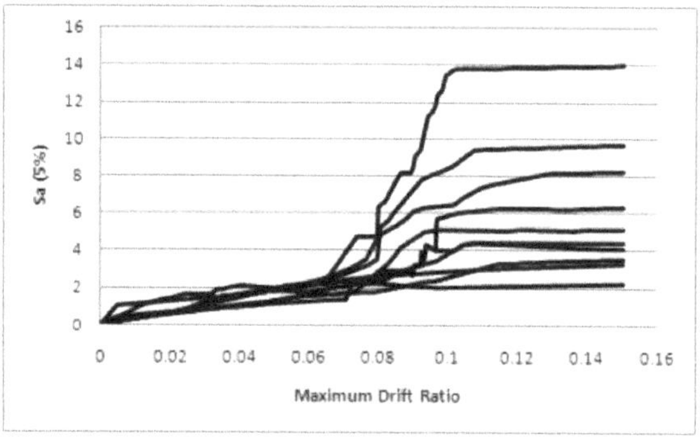

Figure 2: IDA Curves for constrained (lateral support) sheet pile

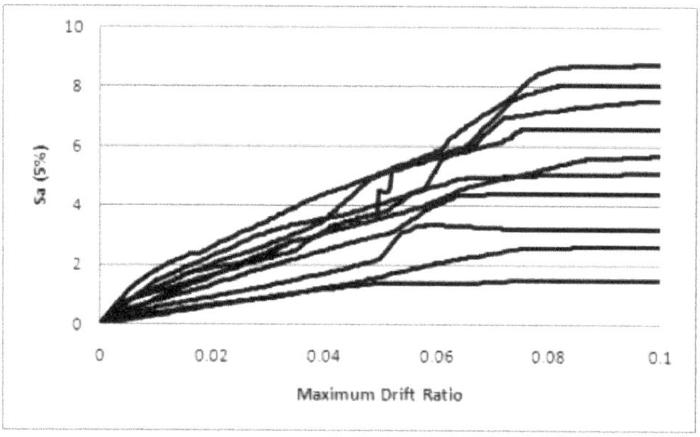

Figure 3: IDA Curves for cantilever (with no lateral support) sheet pile

INTRODUCTION LIMIT STATES FOR SHEET PILE

According to OCDI [2], there are two performance levels for port structures including quays that can be assessed through the following limit state:

- Immediate Occupancy (IO): It's a state limit during which post earthquake structural damages are limited and; lateral and vertically loaded elements still have their pre-quake stiffness and strength.
- Collapse Prevention (CP): This is the state in which structures undergoes sever damages that take it toward thorough collapse; elements have lost their pre-quake stiffnesses and large deformations are found.

Before using IDA curves, limit states shall be defined. For that, according to IDA curves categories and FEMA [6] code, three points of immediate occupancy (IO), Collapse Prevention (CP) and global dynamic instability (GI) are to be identified.

In this study, IO point was selected to be $\theta_{max} = 1\%$, while CP was either $\theta_{max} = 10\%$, or where IDA curve's slope is 20% of elastic portion, whatever comes sooner. Thus, CP is assumed to be where IDA is on the pre straight line ductile area provided that θ_{max} is high enough and the structural model is reliable ($\theta_{max} < 1\%$). Finally, GI is where the curve becomes a straight line and every increment of intensity measure (IM) will result in infinite damage measure (DM). [7]

SUMMARIZATION OF IDA CURVES

For a better understanding of structure's behavior, IDA curves were summarized based on statistical principles. This paper uses 50, 16 and 84 percentages for summarizing. Figures 4 and 5 show the results of summarizing of IDA curves besides limit states for the two types of constrained and cantilever sheet pile quay walls. Numerical results can also be found in tables 2 and 3. One of the other goals of performance based earthquake engineering (PBEE) is calculation of mean annual frequency (MAF) for the limit state. This can be easily done through summarization of the so far obtained results. For that, during this study, a particular integral limit state presented by Cornell [8] was used:

$$\lambda_{LS} = \int_{x=0}^{x=\infty} F_{IM}(x) \left|\frac{d\lambda_{IM}}{dx}\right| dx \tag{2}$$

In the above relatios, λ_{LS} is the MAF corresponding to the limit state or the absolute of IM's risk gradient or first derivative of λ_{LS} - IM risk curve

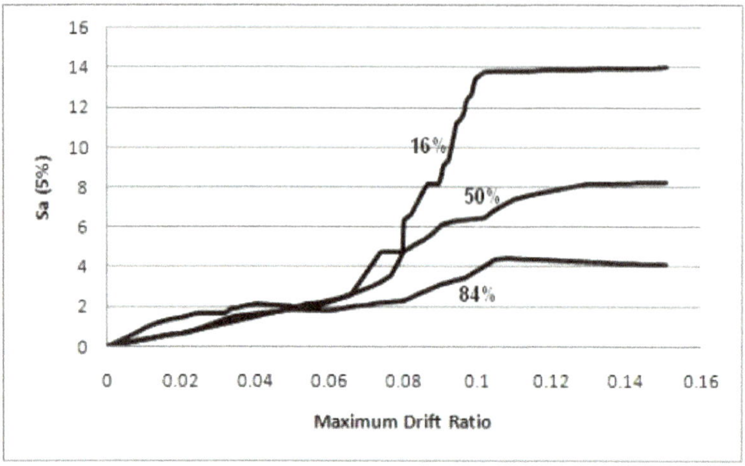

Figure 4: Summarization of IDA curves in the percentages of 16, 50 and 84 for constrained sheet pile

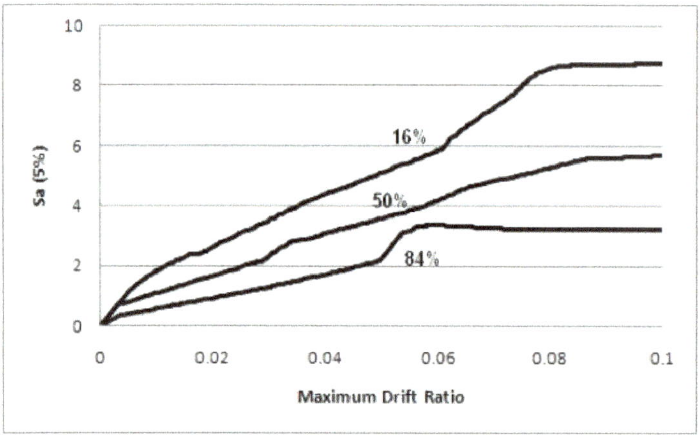

The reverse of λ_{LS} will give the mean time for exceeding a specified state. Different amounts of MAF for various limit states of the constrained quay can be found in table 3 while those of the cantilever one are presented in table 4. Summarization of IDA curves in the percentages of 16, 50 and 84 for cantilever sheet pile

Table 3: Exceeding MAF's and the corresponding return period for the calculated limit state in the constrained sheet pile quay-wall

	IO	CP	GI
MAF	0.00875	0.00089	0.00051
Return Period (year)	114	1124	1961

Table 4: Exceeding MAF's and the corresponding return period for the calculated limit state in the cantilever sheet pile quay-wall

	IO	CP	GI
MAF	0.01075	0.00203	0.00151
Return Period (year)	93	493	662

CONCLUSIONS

According to OCDI, marine structures must be designed for strength level with return period of 100 to 200 years and deformability level with return period of 1000 to 2000 years. Results of incremental dynamic analysis on a sheet pile quay wall for two cantilever and constrained cases shows that the constrained case shall meet the required conditions while the cantilever one will fail to meet the requirements. Obviously the cantilever model will show weak dynamic behavior due to the lack of constrains and this must be taken into account in every analysis.

REFERENCES

1. PIANC (2001), Seismic Design Guidelines for Port Structures
2. The Overseas Coastal Area Development Institute of Japan (OCDI) (2002), Technical Standards and Commentaries for Port and Harbor Facilities in Japan
3. Boulanger RW, Curras CJ, Kutter BL, Wilson DW, Abghari A. (1999) "Seismic soil pile structure interaction experiments and analysis". Journal of Geotechnical and Geoenvironmental Engineering, ASCE, 125(9), 750- 759.
4. Abaqus Manual, ABAQUS/CAE 6.10-1, Dassault Systems Simulia Corp, Providence, RI, USA, 2010
5. Bardet, JP, Tobita T. (2001). "NERA- a computer program for Nonlinear Earthquake site Response Analysis of Layered Soil Deposits" . Department of Civil Engineering, University of Southern California.
6. FEMA. (2000) "Recommended seismic design criteria for new steel moment-frame buildings". Report No. FEMA-350, SAC Joint Venture, Federal Emergency Management Agency, Washington DC.
7. Vamvatsikos D, Jalayer F, Cornell CA. (2003) "Application of Incremental Dynamic Analysis to an RCstructure". Proceedings of the FIB Symposium on Concrete Structures in Seismic Regions, Athens.
8. Shome N, Cornell CA. (1999) "Probabilistic seismic demand analysis of nonlinear structures". Report No. RMS-35, RMS Program, Stanford University, Stanford.

INDEX